The R.A.M.S. Library of Alchemy

Volume 16

Non-Violent Destruction Of the Atom

Hans W. Nintzel

Philip N. Wheeler

R.A.M.S. Publishing Company

Non-Violent Destruction
Of the Atom

Hans W. Nintzel

Philip N. Wheeler

Produced by

Restorers of Alchemical Manuscripts Society

R.A.M.S. Publishing Company

R.A.M.S. Publishing Company
117 Rutherford Lane
Stuarts Draft VA 24477

The R.A.M.S. Library of Alchemy, Volume 16:
Non-Violent Destruction of the Atom
Copyright © 2016 R.A.M.S. Publishing Company

First Edition 2007, "The Betty Story"
Second Edition 2016

ISBN-13 978-1533035684
ISBN-10 1533035687

Image Processing by Philip N. Wheeler

This book is sold for informational purposes only. Neither the
publisher nor the editor shall be held accountable for the use or misuse
of the information in this book.

Printed in the United States of America

Table of Contents

Dedicated to Hans W. Nintzel,

American Alchemist

and

Founder of the

Restorers of Alchemical Manuscripts Society

(R.A.M.S.)

Disclaimer

Liability: The publisher does not warrant or assume any legal liability or responsibility for the accuracy, completeness, or usefulness of any information, apparatus, product, or process disclosed. The publisher makes no representation as to the accuracy or completeness of the contents of this book and specifically disclaims any implied warranty of merchantability or fitness for a particular purpose. No warranty may be created or extended by written sales materials or sales representatives. You should obtain professional consultation where appropriate. The publisher shall not be liable for any loss of profit or other commercial or personal damages, including but not limited to special, incidental, consequential, or other damages.

Introduction
Philip N. Wheeler

I received the first in the series of "The Betty Story" correspondence from Hans Nintzel, dated November 11, 1984. His summaries and additional correspondence from other practicing alchemists soon followed with Hans acting as the communications facilitator. Hans called this "The Betty Story," named after the lady in California who brought the experiments to his attention.

The documents detail an odd pulsating alcohol experiment that some alchemists called the **non-violent destruction of the atom**. This book contains transcripts of that written material. Everything pertinent to the experiments and discussions is included.

This is truly a chronicle of modern alchemists performing and analyzing alchemical processes. This second edition contains added documents that have never before been available to the public.

Thanks for your interest in this unique experiment!

Non-Violent Destruction of the Atom

Letter from Hans Nintzel, Nov. 11, 1984

Alcohol experiment. Put about 30 ml. of 200 proof (or as close as possible[1]) in a 50 ml. Florence flask. Set into a digestive heat where your apparatus is in the East window so it can receive Eastern morning sun light. Above, arrange that green light is shining on the operation (can do several at one time-I am also trying Philosophical Mercury.) Once the alcohol is warm, push in a Saran wrap covered neoprene (or the like) cork. Make it tight. Then, let it "work." Should tinge to yellow, then orange, then red, then thicken, then (happens at night!) a flash of light, silent, then the alcohol, now thick and red as blood, will "pulse" or resonate. A waveform may be seen, 40 pulses per minute across the surface, always in a N-S[2] orientation. This I have seen--ONLY seen, in Betty's Laboratory. Mine is still "working." Do you have any idea of what chemical reaction has taken place? WHY it pulses?

Hans W. Nintzel

---

[1] 191 Proof is as close as possible. -pnw
[2] North-South. -pnw

The Betty Story, Nov. 17, 1984

Hans Nintzel

In the summer of 1984 I called Taylor's Herb shop in Vista and spoke to Keith Taylor about live Celandine herbs. He asked why I wanted them and I explained I was doing some alchemical experiments. He said, "You make the second alchemist I have run into." Naturally I wanted to know who the other was. It was a lady named Betty who lived in San Diego. I called her and we spoke for many hours.

I learned she had been a newspaper reporter sent to do a story on a Dr. Lionel Strong, a cancer researcher. He had developed an anticancer serum from liver nucleosides. Dr. Strong was a member of the Baconion Society (he could decipher the Bacon ciphers) and an alchemist. Betty became interested in both subjects: Bacon and alchemy. Strong showed her some things, especially an experiment he called the **non-violent destruction of the atom**. When he succeeded in doing this, he was so excited he suffered a fatal heart—attack. Betty decided to drop her journalistic career and become an alchemist.

She started by reading Bacon. Deciphering and reading between the lines. She read Eiraneus

Philalethes and became convinced that he was, in reality, Sir Francis Bacon. The latter supposedly wrote the plays of Shakespeare's, Burton's "Anatomy of Melancholy" and was also the supposed author of the Christopher Marlowe materials (who was also thought to have written Shakespeare's plays). Philalethes first name is one letter away from the Latin word for "hedgehog" which is a symbol used to denote Bacon. This and other things convinced her that Philalethes was Bacon and thus all his writings were to be avidly read and pondered.

The first experiment was to take fifty (50) ml. flasks and place in them about 30 ml. of pure, 200 proof, grain alcohol. These she set on a hot plate (food grade device) and set this into an Eastern window so it would get the morning Eastern sunlight. Over the set-up she hung a green light. She set the hotplate to a low or digestive heat. Then she allowed it to "work." After time (a few weeks) the alcohol took on a yellow tinge. Then it went to orange and finally a deep red. At some point the "non-violent destruction of the atom" took place, at least according to Betty. That is, a flash of light occurred in the early morning hours and the alcohol, which had now coagulated to a chocolate syrup-like consistency, began to *PULSE*! It was measured at 40 beats per minutes and operated, always, in a North—

South direction.

Since I went to San Diego to meet her, and see it, I can describe it. In the window, in the little flask, was a thick, dark red matter. It would send a sort of waveform across the surface of the fluid every 1½ seconds or thereabouts. There were no tricks as I very carefully examined it. Being a former stage magician, in a minor way, I looked for "gimmicks." I could find none. I also spent a day and night with Betty talking about alchemy. I was the first alchemist, other than Dr. Strong, she had known. Of course, she had by now met Dan from Australia along with Art and Jim from Los Angeles, who at my suggestion went to visit her

We feel she has far too much intuitive knowledge on alchemy than can be explained by the word "intuition" or by chance. We feel that she is the reincarnation of a devotee of Sir Francis Bacon!

Her next experiment concerned alcohol as well. What she did was to make the fixed tincture of antimony as revealed by Basil Valentine in his book **Triumphal Chariot of Antimony**[3]. Once she had the gum, she macerated it with pure, grain alcohol. Then she distilled the alcohol off the antimonial gum. She

---

[3] Volume 2 in The R.A.M.S. Library of Alchemy

called this "AA" or, antimony purified alcohol. But she had no idea what it was for. She was acting on pure intuition!

I told my friend Kurt, in Salt Lake City, about her. He felt impelled to call her and told her she could use the "AA" as medicine. The dosage was one drop per day in a little warm wine. Or, if she could handle it, twice a day. Betty had other experiments but the main point of this paper is to describe the effects that the "AA" had on her.

She wrote to me about those effects. Prior to that, Jim called to tell me about the same thing. He mentioned, and she did not, that she had a heart condition that prevented her from walking more than, oh, half I block before she had to rest. After taking the AA for twenty (20) days, she found she could walk 5-6 blocks. Jim started taking the AA and found that his energy levels quadrupled!

Now, in Betty's letter, she related that after starting the regimen of one drop of AA in an ounce of warm wine, morning and evening, she experienced a cessation of chest pains she usually had and which she controlled by taking the drug Vasodyline, a vascular dilator. After the first two days of taking AA she was able to stop her daily routine of

Vasodyline (10 units three times a day) and had no further pains in her chest. While she felt much better physically and mentally, she experienced an unusual side effect. That is, every 3—4 days, around 11:00 AM, she became very sleepy and HAD to nap. The 'nap' sometimes lasted all day! She feels though, that this will pass in time. That perhaps it is an adjustment being made by the body.

She is also a diabetic. She takes insulin but has decreased the dosage from 44 units a day to 32. Her urine output, previously 8 times per day, is now a normal once in the morning and once in the evening, in terms of frequency. She has noted that the *urine* contains a clear and viscous substance that coats the sides of the commode but will float off in thin coagulated strings. Again, she feels that these are wastes that the body can now eliminate due to the AA.

Finally, she has had a blocked Eustachian tube that caused a hearing loss. Since taking the AA, the Eustachian tube has become unblocked and her hearing, it part, has been restored. Amazing!

My opinion is that she has been able to make a sort of intuitive "break-through" aided and abetted by various folk. It would seem that a fruitful line of research into antimony purified alcohol is open, and one that has shown great promise medicinally. Now, the alcohol that is pulsing or "resonating," this too may have promise as a medicine. Also the fruits of some of her other alcohol experiments. She is a true pioneer in this area as is shown by her willingness to test the products on the ultimate subject, herself.

I urge you, my fellow alchemists who read this, to direct some efforts along these lines. With several researchers investigating the various properties of the alcohol preparations, new facts will be revealed to enable all of our fellow men to live in greater health.

In a subsequent conversation with Jim, he indicated that in HIS ingestion of the AA, he noticed, as indicated, a great deal more energy. Also, he had been allergic to coffee. It caused bowel pains and other physical discomforts. While taking the AA, he could drink as much coffee as he cared to without discomfort.

Further, he contacted Kurt and discussed the experiment on alcohol (AA) with him (Kurt). Kurt told him that antimony is indeed the wolf of poison in the spirit. That is, it will absorb all poisonous matter from a spirit cohobated upon it.

Kurt had made the dry distillation of lead operation and obtained the Golden Water. He then macerated this on stibnite ore for about one month. This, he claimed, freed the spirit from the poisons. He then took this antimony purified Golden Water and cohobated it upon the dead head (feces) remaining after the dry distillation. This caput mortuum contains, to be sure, salt of lead. This Golden Water was then distilled off of the dead head. Then cohobated back on and then re-distilled, etc.

This was repeated until Kurt saw crystals (the salt) in the neck of his flask. What was happening is that the salt was being chemically "married" to the spirit. Kurt then took this liquid and placing it in a suitable vessel, set it outside to absorb the various forces of nature i.e., Sunshine, cold, starlight, moonlight, heat, fog, snow and so forth.

Kurt said that in a year's time, while it was thus digesting, he saw CLEARLY in the flask the steno of the zodiac, the letters of the alphabet. "God's alphabet" as he called it. Then the colors started to show. The *Cauda Pavonis*. A sign of putrefaction. This is where he is with this, at this time. (Nov. 20, 1984) He will await further changes and then...

try some on himself.

Jim felt that the liquid, in a sealed flask, in a sand bath, would putrefy more rapidly. Well, could be. Again, here is fertile ground for experimentation.

Letter from Hans, November 28, 1984

I am enclosing a little "dossier" I wrote concerning Betty and the Antimony purified Alcohol. It will give you the "big picture." What is interesting is that NOW my experiment, the GRAIN alcohol, has turned to a deep tan color in a day or two from a pale straw color over about 2 months being in the heat. SO. I am excited about this. The GRAPE alcohol is not doing ANYthing. Hummmm. I would suggest you use, in tandem, a small flask of GRAIN alcohol. I used Everclear from the liquor store. And I was too lazy to rectify it. However, it IS 96°proof so...

My flasks are in a round tray of sand sitting atop a 6" diameter hot plate that has only three settings. Hot, Medium, and "low." I can hold some fingers on it for a few moments before it gets too uncomfortable. The green light is a garden-type light bought in the hardware store. A flood light with a green glass front. (Betty uses an ordinary bulb shining through a piece of green plastic sheet!) The "format" is not so germane as is the fact of a "green ray" being used. The whole apparatus, including a tripod, sits in a east window. Not getting TOO much sunlight as there hasn't BEEN that much the last few weeks. So, sun, green lamp and hotplate provide the "energy."

I would like to see you and Bears think about what is going on in the alcohol. And why the antimony "purifies" it and what the ♁[4] is removing (or adding!). I believe this is the verge of a "breakthrough." Your training in the formal methods of chemistry can be of immeasurable value in putting this experiment in a "scientific light." So, THINK!

Well my lad, all for now. I DID write to Bears about the translation, let's see what he responds.

Hans W. Nintzel

---

[4] The alchemical symbol for Antimony. -pnw

Journal: A Pulsing Alcohol Experiment

Hans Nintzel

This experiment was conducted to see if ordinary grain alcohol, when subjected to certain influences (heat, green light and sunlight) would react as described by Betty McKaig. That is, turn color from yellow through brown-orange and to a dark red. Then, after an uncertain time period, commence to beat or pulse.

9—09-84 Set up a hotplate with a dish of sand in an Eastern facing window. Arranged via a tripod, for a green 100 watt spotlight to shine on any flasks sitting in the sand. At this time, used three flasks: a 300 ml Kjeldahl containing 150 ml of alcohol; a subliming device with 100 ml of alcohol and a 100 ml flask with 50 ml of alcohol. All of the alcohol I had distilled from burgundy wine using a Vigeraux column. The flasks were sealed with neoprene stoppers wrapped with "Saran-Wrap." At 1:00 pm the sand bath was turned on as was the green "ray."

9—11-84 Added to the three flasks a small 50 ml ground glass bottle with 25 ml. of wine alcohol in it.

9-15-84 The subliming device, a pair of wide mouth Florence type flasks, had all the alcohol evaporate out. Replaced it with a 125 ml. Florence flask with 50 ml of wine alcohol. Now "luted" all flasks with G.E. "Glue and Seal." Returned all to sandbath. The spotlight had burnt out and I replaced it with a G.E. "miser" type green spotlight. Gets 100 watts of light with 85 watts of power. - The little bottle also lost all alcohol.

9—19—84 Alcohol in the 125 ml Florence flask turning yellow, but seems to be evaporating. The color is a medium yellow.

9—26—84 Alcohol in 125 ml almost totally evaporated! The residual has fat—like globules in it that are quite red in color. In a visit to Betty she gave me some "antimony purified alcohol" which I will start to take soon. She gave me some small 50 ml flasks and some absolute alcohol. I poured 25 ml of her absolute alcohol (straight grain alcohol) into one of the 50 ml flasks. Filled (after cleaning) the 125 ml flask with 35 ml of her absolute alcohol and 15 ml of her antimony purified alcohol, to make a total of 50 ml of alcohol. Luted both with G.E. Silicone Sealer.

9—29—84 The 50 ml flask with 25 ml of abs. alcohol has entirely evaporated. The 125 ml flask had a reddish streak running down its side. Resealed 125 ml flask with nail (clear) polish and more silicon sealer. (stripped off first luting) . Took another small 50 ml flask and put in it 30 ml of "Everclear" (grain) alcohol which is 190 proof. Sealed with fiber tape, heavy nail polish layers (as suggested by Betty). AND luted with Silicon white) Sealer. Did the same with a 50 ml flask filled with 30 ml of Spirit of Lead. (obtained from lead acetate and extracted from the Golden Water). The latter was, over time, tinged again to a yellow color. Also added to this flask some volatile salt of Saturn which I flushed from the condensers after dry distilling the acetate. Finally added a small (ounce?) bottle to sand bath containing 10 ml. wine alcohol.

10—01—84 Alcohols STILL seem to be evaporating. The spirit of Lead has tinged to an orange color. Sealed all flasks now with hot wax.

10—10—84 The little bottles and the flasks have had ALL fluid evaporate. The spirit of lead left a thick red residue. I refilled (after cleaning) the three small flasks. One with more spirit of Lead (but no salts). One with grain alcohol and one with wine alcohol. I warmed the liquid and then inserted Saranwrap covered Neoprene stoppers. Let them cool then dipped them into hot wax. Place asbestos mats in sandbath.

10—10—84 The purpose of the asbestos mat was to reduce heat a little. My hotplate has only three settings, low, medium and high. The stopper on the acetone kept popping so I felt heat was a bit high.

10-26—84 The acetone has tinged an orange—red with some dark viscous matter floating in it. The grain alcohol has tinged to a straw yellow. The two larger flasks (wine alcohol) show no color whatsoever.

11—06—84 No changes in color. The last few days, overcast, little sun.

11—18—84 Ditto.

11—25—84 The 50 ml or grain alcohol has evaporated somewhat and is now a deep yellow tending to brown color. No other changes.

12—02—84 Removed the 300 ml wine alcohol flask as there seems to be no changes whatsoever. Inserted a 100 ml flask (Florence with a flat bottom) containing 60 nil of grain ("Everclear") alcohol.

CHRISTMAS Occasional observations indicated no changes taking place. However, the 50 ml of grain alcohol somewhat darker and reduced in volume (or is it just evaporating?). Acetone darker, reduced in vol.

NEW YEARS DAY Comments are the same as for Christmas.

01—14-85 The Green bulb burned out and I replaced it, A GE Miser. Now, the one I replaced was actually the 150 watt spotlight. The FIRST one was a GE Miser 85/100. The 150 watt spotlight lasted 5 weeks or so and cost three times as much as the GE.

The acetone is in two forms now. A light brown fluid floating atop a very dark brown, tending to black, substance on the bottom of flask. The grain alcohol is a medium brown and evaporating more! None of the other alcohols show ANY changes.

01—18—85 The last few days have seen plenty of sunlight. No visible changes in the flasks.

01—30—85 The 50 ml of grain alcohol is reduced to maybe 10 ml. Its color is a deep brown. The 100 ml flask of grain alcohol is seeming to "circulate," drops forming on the neck and then precipitating down into the body of the flask. But no color changes. The acetone is deep reddish, dark brown with what seems to be flakes of matter in it. The wine alcohol shows no changes whatsoever. I must point out that these flasks 50 and 125 ml, have longer then usual necks (I wanted then to circulate).

02—05—85 Very little grain alcohol left but the color is quite dark. The acetone is also a bit darker, it seems. No change in wine alcohol.

02—09—85 A fateful day!!! The grain alcohol was so scanty the last few days, it did not even cover the entire bottom of the flask. Due to a slight bulge in the center of the flask, the remaining alcohol, a deep, dark—red color, "rolled" to the edges of the flask. I had decided to start anew: However, as I looked at the grain alcohol (no other changes in any of the other flasks) I was startled to see what seemed to be movement. I carefully watched and LO, it was indeed PULSING! It was a sort of jerky movement, very rapid, and then it slowed down to a sort of cyclical heat. Then speeded up a little and this seems to be the pattern. As best as can be discerned, there being little fluid left and this motion being on one "end" of the crescent of dark red alcohol, the beat seems to start North and move South. There seems to be 45 beats per minute—counting fast and slow ones. IT PULSES!!!

02—10—85 Two witnesses observed the pulsing. However, this day, the pulsing stopped! The liquid more viscous, lesser in volume than before!! But our Father has permitted me to witness this. Hopefully, we will be able to understand its use and Purpose.

Letter from Hans, Dec. 19, 1984

I am enclosing some papers from the pen of Arthur
Fehres. Arthur lives in Australia and is a fellow
alchemist who has done a fair amount of independent
research. He is commenting on the "Antimony purified
Alcohol," or AA, that our friend Betty came up with
in San Diego.

I am hoping that these remarks will cause YOU to put
on your thinking cap and perhaps suggest some areas
of research or experimentation that might be useful.
Or, provide any data from the literature or of your
own that deals with the use of such material as a
medicine. Or your ideas on what is happening. You,
to whom I write, have fruitful imaginations and
hopefully, we will obtain an outpouring of useful
ideas and data that will take us further along in a
quest for a valuable medicine.

Please communicate to me any information,
suggestions, ideas, citations from old (or modern)
literature that might add to a body of knowledge on
this subject.

                    Hans W. Nintzel

Letter from Arthur Fehres, December 1984.

In response to the request to direct some efforts towards investigating various properties of alcohol preparations as described in 'THE BETTY STORY' I would like to contribute the following.

REGARDING ALCOHOL

Alcohol is the volatile mercury or spirit of the plant kingdom. It is flammable, expanding and heating and therefore energizing, but by itself has no medicinal properties. It is the vehicle of the life force, freed during the process of fermentation, which is absorbed by and within the medium of water. In sugar and starch this life force is fixed. Once the energy is freed and one imbibes oneself with it, the body becomes more and more spiritualized resulting in one's consciousness becoming more and more subdued and the mind or computer eventually goes berserk. For this reason, freed energy (i.e. not fixed, or separated from the sulphur and salt which held it 'captive') is actually dangerous to organized life forms.

Each cell of our body also has its own consciousness. It has been proved that cancer cells feed on alcohol, whereas normal cells feed on sugar. In my understanding, cancer occurs when given—off life energy from dying cells cannot leave the area quickly enough. It is then absorbed by the lymph fluid, in which every cell bathes, forming a kind of alcohol. Healthier cells in that area consequently become intoxicated. Their reigning consciousness gets out of control and an inherent survival mechanism is triggered off within one of the stronger or strongest cells resulting in growth independently from brain control. It has been proved that excess alcohol increases the death rate (destruction) of nerve cells.

Not in one Materia Medica is alcohol, no matter how pure, listed as a medicine. Alcohol should be looked upon, firstly as an extraction agent or menstruum to obtain alchemical sulphur; secondly as a magnet for this sulphur and thirdly an ideal medium to preserve medication, for which reason it is extensively used in homeopathic triturations (potenitized dilutions).

REGARDING THE FIRST EXPERIMENT

This one is most fascinating and I do thank Betty for sharing it with fellow alchemists. I have not performed this particular experiment (I intend to soon) but I have become aware of the importance of Eastern sunlight. For instance, I have exposed volatilized antimony tincture, but never for weeks on end.

Also, for some years now, I have occasionally been watching the sun rise, on clear days, to absorb this special light through the pupils. Seconds before the sun becomes visible at the horizon, one fixes one's eyes (staring) on the brightest spot. When the star appears the light is very bright, but the growing golden orb is somehow blotted out by a light grey of the same size, leaving the edges or corona exposed, not unlike an eclipse. As the sun rises the light intensifies, the blotting out weakens and becomes less effective. From experience I know it is very safe for me to stare at the star until twice the disc's diameter is between the horizon and the disc, which takes 2 to 3 minutes. If kept up for longer, which is not recommendable, the eyes get a little sore. I have good eye-sight and no need for glasses. <u>Maybe</u> the proverb 'Early to bed and early to (the sun's) rise, makes a man healthy, wealthy and wise,' should be understood with what is in between brackets!

Early morning sunlight, which is always from an Easterly direction, is more abundant in ultraviolet rays. These rays are most penetrative and therefore I think that they carry the essence of the sun in a purer form than, for example, infrared, which has deteriorated into heat.

In the experiment, I think that the grain alcohol (S.V. should be tried too) attracts, absorbs and stores this pure solar essence, in which the latter accumulates and becomes more and more concentrated until male and female (solar sulphur and mercury) have balanced each other. I think, it is at this point that the flash occurs, indicating the 'birth' of the third. Of course, all this is speculation, a hypothesis for contemplation, but to me it seems more feasible than "the non-violent destruction of the atom." To me, the pulse or wave form indicates life rather than destruction. Also, photons (light) are released when an excited atom drops back to a lower energy level without necessarily being destroyed.

It is interesting to note, that Cockren's Golden Water (GW) also turns red, the longer it is kept. More volatile fractions of a light-yellow eventually turn a deep yellow and those fractions which look clear take on a faint yellow tinge in time. Once the volatile mercury has been separated from its sulphur it stays as clear as alcohol. The change from yellow to red I understand as being due to degradation and subtle densification of sulphur as a result of heat, subtle or otherwise, and/or the effect of light upon it.

However, 200 proof grain alcohol is devoid of grain sulphur. The yellowing seems to be proof, that it has its origin in 'the East.'

REGARDING "AA"

It follows from what I wrote regarding alcohol, that "AA" or 'antimony purified alcohol' is a misnomer. The maceration of the gum allows the alcohol to absorb the most volatile sulphur (the virtue), which the alcohol carries with it during distillation. The potency of this 'medicated alcohol' depends on how much alcohol is used for maceration, not to mention time, temperature and other factors.

REGARDING POTENCY AND DOSAGE

In view of research, to facilitate clinical evaluation of products of this nature, it is essential to use standardized potencies. Having practiced homeopathy as one of the healing arts, I can recommend the following method AFTER THE FINAL EXTRACT HAS BEEN MADE WITH ALCOHOL:

a. Gently distill off all alcohol, which then yields the gum.
b. Note the exact weight of the flask plus contents <u>together</u>.
c. Redissolve as much of the gum with as little

warm) alcohol as conveniently possible.

d.   Weigh the flask plus remains (when dry) and subtract from the weight obtained before.

This gives you the exact weight of what is in your alcohol and at the same time you have rectified your subject. If this weight is 1 gram, then this 1 gram I consider and call 'solid extract.'

1 gram of solid extract makes 10 milliliters of fluid extract.

10 ml of fluid extract makes 100 ml of tincture.

100 ml of tincture makes 1000 ml or 1 liter of 1x

1 liter of 1x makes 10 liter of 2x.

1 liter of 2x makes 10 liter or 3x; 10 l of 2x makes 100 l of 3x

So, I gram of solid extract makes 1000 liter of 4x, etc.

e.   Adjust your product to the desired potency.

The dosage is determined and measured it drops. Throughout the ages the choice of potency has always

been a matter of trial and error, more or less
relying on intuition as a guide. However, rapid
growth in ultra-sensitive electronic computer
technology, has produced the Theratest. Research
indicates that each individual requires different
potencies for optimum benefit. The Theratest
indicates which potency it is. A friend of mine is a
pioneer in this field. You will hear more about that
when sufficient data have become available.

REGARDING KURT'S EXPERIMENT

Some points worth considering.

1. It should be understood, that crude antimony is
saturated with all the poison it can hold;

2. The acid in the GW[5] dissolves some of the
antimony and mingles with it;

3. During maceration of GW putrefaction takes place;

4. What appears to be a Cauda Pavonis[6] is not always
a sign of putrefaction;

5. GW IS body, soul, and spirit.

---

[5] Golden Water. -pnw
[6] Peacock's Tail, a reference to its many colors. -pnw

40

Letter from Jim Woolsey, Dec. 26, 1984

California

Dear Hans:

Thanks for the Arthur Fehres Papers. I am interested in GL, because I have spent much of my time mentally "seeing" connections which, in my current opinion, only prove the ability of the mind to create worlds, but not necessarily to see what is workable. Rephrased, scientology-wise we might say that each being has the power to create particles which form into masses, the material world. The so-called material world about us is the sum total of creations of all individual spiritual units or beings; and what we culturally agree upon, our group reality, is sort of the average, or what is common in all our world views. Reality is Agreement. Enough of us say it is so, then for practical working purposes, it is so (at least for now) and we call it REAL. Its permanence depends on continued agreement.

Each being also has his own personal universe, wherein whatever he wishes to be, instantly exists. What you imagine instantly appears clothed in mental matter. To get it clothed in heavier more physical matter is indeed the work of the Magician. You probably understand what I say quite well, even if the words are different.

This being so, seeing all these intricate connections may only display the mind's ability to create. But it may not reveal a path or technique workable with the agreed upon reality of this greater world. If you want to spin your wheels on the mental plane, great. As you have experienced, we can make the alchemical texts fit any current theory we hold, even if it yields no results in the lab. For me, the AA theory gives the most comprehensive translation of alchemical symbolism so far.

Of course, as I say, I love to play GL myself, and can usually select details from my world which show how special I am. For instance, the path (freeway) to my home has these off ramps: Montrose (Rose on the mountain, the Rosicrucian symbol of attainment), La Crescenta (the Moon, and counterpart to Sunland, wherein I reside); Ramsdell (the dell of the Ram, which Betty says is a key word meaning Antimony.) Lowell (= low well, a deep hole, or a black hole, into which matter is compressed as in Betty's AA work, or in the black hole in Cygnus where also the Great American Nebula is found, and various supernovae have shown, as in Betty's starmap); Tujunga (Indian for "big winds," i.e., the spirit, or mercury, the child born in "our sea," the atmosphere, "the wind carried it in its belly"). And of course I live on Day St., which is the intensification of the Solar influence found in Sunland, so where else would a budding adept live? Fun, but maybe no gold ring.

Enough of this. I would like more info on alcohol feeding cancer cells. I can understand how undischarged energy might upset the cellular equilibrium and cause cancer. As you know, I took Reichian therapy from Regardie for 4—5 years. In this theory, there needs to be a proper balance between the inflow and discharge of energy. Body armoring (due to patterns of fear and anger and pain impressed upon the body) disrupt the natural circulation of life force, orgone, and cause some parts, organs, cells, to be either deprived of life force, and they wither, or over saturated, causing tension, anxiety, and some physical malaise.

The cure is to restore balance and get those flows going. My experience was that we do not discharge enough; we hold in our waste products, be they energy or matter. To equalize the system, the therapist introduces MORE FREE energy into the system, on a gradual basis. This is done thru breathing. The patient has to forcibly <u>expel</u> the energy also. The body gets quite heated up, with strong energy flows, and blockages are broken down. There is a gradient loosening up of the patterns of armoring. Reich could get cancer patients to go into remission with his technique, but the cancer could reassert itself, and would do so until the entire blocking pattern was dissolved out by repetitive rebalancing (i.e. cohobations of energy upon matter).

So Free energy is not the problem, the problem is improper discharge of excess energy, or lack of discharge. The energy is held, becoming a mass, matter, and the body hardens, gets more fixed, more mineralized, however you want to say it, and dies.

Even breathing is continual influx of free energy, as I see it. Pranayama is fundamental to most systems of liberation, which attempt to break up the holding patterns of the spirit, which tie it to the earth, or rebirth. These patterns, *skandas*, aggregates, or mental pictures, facsimiles are the baggage that chain us to rebirth. But I digress.

A healer heals by inflowing free energy into a stagnant or stuck system, restoring equilibrium.

So free energy is not the problem in my opinion. But

we have to regulate the resultant cleansing process so we throw off the released junk. Or we die in our own filth.

But Arthur, who has **not** seen Betty's flasks, misses two important points:

a) Not only does it absorb light and become thicker, but it starts pulsating, a living beating heart (also depicted in Fulcanelli's color plate). This to me clearly shows the life force of the sun becoming fixed into an organism, the perfect miniature recapitulation of evolution and the spirit to matter—life form trip

b) The purified alcohol, AA can breed life, and little microscopic critters scamper around in it. Alcohol by itself kills life and is a disinfectant. So this AA is no longer alcohol as commonly known.

For all of his GL and work on butter of antimony, I do not think Arthur can show anything as unusual as these two points. Not to mention the incredible shrinking and disappearing AA, which none of us have seen, but I give Betty a lot of credence.

As to medicinal uses, I now have more people trying it. One friend is very sensitive, says it immediately energized her, but then she got very drowsy, so stopped. You remember this happened to Betty. I have been rather sleepy lately myself. I will keep you posted.

I also have some GW[7] on the hot plate and it darkened quite fast. There is a grayish ring left on the flask. Did you get this?

I am thinking that we should use AA for the Junius work also, nicht wahr? An activated spirit should hasten the putrefaction and elevation of the salt.

My brain is now addled, no doubt due to excess free energy, and I must end......

Jim

P.S. Such correspondences as "crow" and "crown" may exist in English, but certainly did not in the original languages used, such as Latin for the Turba. This is indeed mental masturbation (= maceration = musheration), which, while it lights up the mind and senses with pleasure, gives birth to nothing.

_____

[7] Cochren's Golden Water. -pnw

P.P.S.   Danae, who is my better half, tells me
that while I have displayed my wit at poor Arthur's
expense, I have not fairly shown appreciation for
his good side. This is no doubt true. He is one of
the admirable students of alchemy. He is certainly
no vainer than me. You, as the wise and sensible
middleman and keeper of the records, stand aside and
collect these diatribes and fulminations. But I feel
that I can send you my overwrought invective, and
you will know to enjoy it for its literary worth,
rather than taking it to mean Jim is mean in
praise, affection, or appreciation of other
students.(!)

A Passage from Schroeder on Antimony Glass

N. Aus diesem Kalck kan man ein gutes schweißtreibendes Mittel bereiten/ das vor der Pest bewahret/wann mans weiter in einem ☩ glühet / daß aus einem weissen/ ein gelblechtes Pulver werde. Dos.gr. iij. in Wein infundiret / man giebts mit Ringelblumen-Safft.

Man glühet diesen Kalck entweder weiters in einem Tiegel vor sich/ daß es ein gelblechtes Pulver werde/ oder man thut ana C.C. ustum. darzu/und lästs im Feuer calciniren. Und alsdann besitzet er neben seiner schweißtreibenden Krafft auch eine styptische / dahero kommet er auch in des Hr. D. Joh. Michaelis Rothruhr-Pulver.

Wann gemelter Kalck in einem Tiegel wol geflossen (welches man mit einem Eisen erfahren kan) so giesset man ihn auf eine steinerne Tafel/oder Küpfern Becken/daß er zum Glaß werde. Ist es noch schwartz und dunckel/ so muß mans wieder giessen bis es klar und jachzinthen Farb wird; In welcher Schmeltzung Matthiolus was vom rohen Antimon. darzu thut. Allein möchte es davon nur dunckeler werden. bes. Basil. im Triumphwagen.

N.1. Damit es desto eher calciniret werde als kan man halb so viel gemein Saltz darzu thun/ und solches hernacher durch ein siedend Wasser / und dann ein wenig destillirten Essig abwaschen.

2. Damit es desto eher schmeltze/ so kanst du zu ℥j. Antymonii, ʒß. Borax thun / und zwar sonder einige

Gefahr/ weil derselbe meistens wieder verrauchet/ ja man kan ihn auch ohne Schaden ʒß. schwehr gebrauchen.

3. Etliche erwehlen auch zur Bereitung der Gläser/ eine gewisse Zeit/ wann nemlichen ░░░░░░░░░ zu ░░░ (als wässerigen Zeichen)stehen.

N. Wann das Glaß annoch dunckel wäre/ müste man die calcination und Giessung wiederholen/ bis es Jachzinthenfarb würde.

Wann man Saltz und Borrax darzu thut / so wird es nicht rein/ noch auch so durchscheinend und Jachzinthenfarb / sondern bleich-gelb. ░░░░░░░░░░░░░░ vornehmen/ welches man auch in Bereitung des Reguli beobachten soll.

Wann man in der vitrification, das geflossene Antimon. mit Eisen herauszieht / selbes zerstösset/ und wieder schmeltzet/so bekommet man ein dunckelbraunes vitrum, wegen Beymischung der Eisen-Schlaggen.

4. Dieser vitrification kan auch folgende beygezehlet werden / als worinnen das Antimon. gar leicht durchs Feuer in ein nicht durchsichtiges Glaß verwandelt wird: Man läst nemlichen Antimon. in einen Tiegel fliessen/ dann stösset man eiserne Stäblein drein / und schläget diß/was daran klebet/ herab/ auf diese Weiß ziehet man alles Antimon. heraus. Dieses zerstösset man/ und lästs wieder fliessen/ ziehets heraus/ wie zuvor/ wiederholet es so offt/ bis das Antimon. gepülvert
eine

48

Favorable Constellation for the Work

Feb. 20, 1985

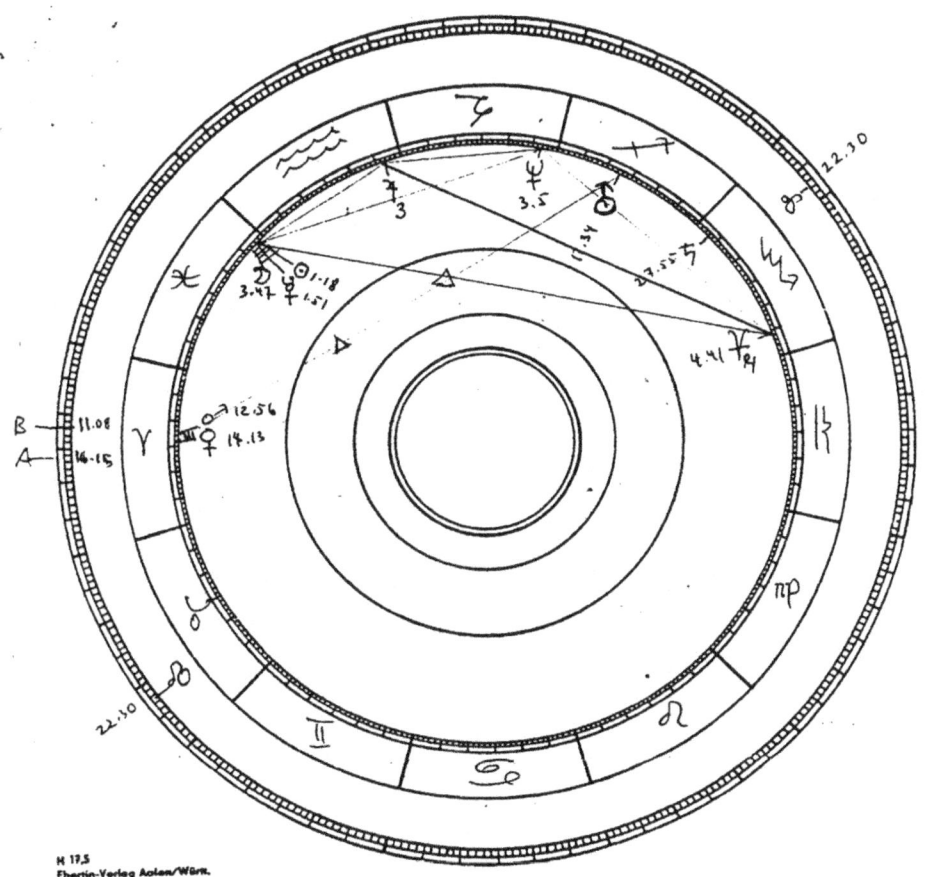

Letter from Hans, 1985

Enclosed is the latest of the "Betty Saga" from Betty herself. As you can see, your wish is coming true as to her sending some samples of the alcohol *"goodies"* to Bears.

The class. Sigh. I had a mailing list from ESSENTIA. Apparently it is WAY out of date, or, the majority of our brethren and sistren live the life of gypsies. I sent out about 150 of the class letters, got 50 returned as "Address Unknown," "Expired Change Order," "Deceased," "Refused" and the like. Got three no thank—you letters, yours makes it four, and 4—5 people who sent a check (or will help finance) for the class. At this point I am discouraged and think I will have to scrub the deal. Too bad.

Evaporation and contamination. I too wrapped my Neoprene (inert) stopper with Saran—Wrap. Couldn't BE any contamination or tincture—of—cork. Also, before using the clear nail polish, I wrapped it with filament tape. One (the one that finally worked) I dipped into hot wax. The others, I used G.E. white silicon glue (stuff). I really don't think we can attribute the reaction to contaminants. Maybe enzymes. Maybe formation of polysaccharide strings. Now Betty simply pushed in a neoprene stopper. Used no sealant (luting) for the color change, pulsing experiment. On the other hand, "proper contaminants" may be common bacteria!

Um, on evaporation, I am having two 50 ml quartz flasks with stoppers made. Quartz, as it will allow more ultra violet in and as this is a tighter seal. Hopefully, this will solve the evaporation problem. One could also heat the vessel PRIOR to insertion of the stopper, heat it a little hotter than the environment in which it will be. Cool it AFTER putting in the stopper. The vacuum thus formed will give a better seal. Yes?

Well, my man, all for now. Hope all is well and if you come to the mid-west, it will be close to Dallas!

Hans W. Nintzel

Letter from Hans, Nov. 15, 1984

The alcohol situation has taken an unusual turn, as I will describe. But first, what do you mean by "resonate" when you indicated some organic experiments must be first *resonated* before they will work. The alcohol I described was pulsing in that a wave form went across the surface. It did this of its own accord and not by any catalyst or outside influence. So let me know what YOU mean by resonate. O.K.?

The lady in California indicates she is ingesting "antimony purified alcohol." What she did is make the gum from glass of antimony (fixed tincture) and macerated the gum with ethyl alcohol. After a tincture was drawn, she distilled the alcohol. Not sure if this was repeated but let's say she did it three times. Now she has what she call, "Antimony purified Alcohol."

She was advised by a friend of mine in Salt Lake City (who called her, when I told him about what she had) and congratulated her and told her she need not do anything further but could now use this alcohol medicinally. He suggested a drop or two per day, in wine, for 20 days, then lay off for ten days. (To allow the body to assimilate and react to the "purging"). This she did. She found that, despite a heart condition which prohibited her from walking more than a ½ block, she could now walk five blocks. A blocked Eustachian tube, causing partial deafness, opened restoring a good part of her hearing. Now, Jim, upon hearing all this, makes up a batch and takes it. He swears he has lost some weight (as was predicted) and has a tremendous amount of new-found energy, SO!!

I am now looking for tracts on alcohol! It seems that this substance, simple though it seems, has significant medicinal value that should be investigated. I plan on making my own, or, distilling the alcohol off the fixed tincture of Antimony I bought at ParaLab. The Antimony, is the "wolf of poison," absorbs it. So, one should NOT ingest the antimony part, since it contains the poison. As, we suspect, does the ParaLab product. The "Master Enzyme of the World" is now in the alcohol, transferred from the antimony. It is, seemingly, a medicine of great value. And, with some more "work" this alcohol may become a panacea!!!

Any ideas? Comments? Works dealing with Antimony? We may have something here and you and Bears could possibly supply some academic or scientific impetus and explanation. But so long as it WORKS!

                              Hans W. Nintzel

Letter from Philip Wheeler, Jan. 6, 1985

Columbia, MD

Regarding the Betty Experiments, the Arthur B. Fehres comments, and your last few letters, I will now try to answer some of your questions, advance a few theories, etc.

RESONATE:

In my letter early in November, I mentioned the "resonating" of certain organic experiments. I was referring to the practice of placing a container of reactants near a source of ultrasonic sound. This is, of course, not what you meant by resonate, which was made clear by your communications subsequent to Nov 15. One possible explanation for the waveform on the alcohol would be that an oscillating chemical reaction is taking place. There was an interesting article on oscillating chemical reactions in the March 1983 issue of Scientific American. The reaction could involve polarized molecules. This is only a crude hypothesis, but it could explain the north—south direction of the waveform if the polarized molecules were being affected by the earth's magnetic field.

BETTY EXPERIMENT #1:

I started an experiment of this type on Dec. 2 (using liquor store grain alcohol, once rectified). I have a 100 watt green floodlight about 5 inches above the flask. The flask itself is sealed with a Teflon stopper, which should preclude any possibility of reactions involving stopper material (Teflon is completely inert at temperatures below about 250 degrees C). It hasn't gotten too much sunlight; the spring and summer will be better for that aspect. So far, I cannot detect any color change. The temperature of the sand near the flask is kept at about 75 degrees C. What we need to investigate is the range of possible photochemical reactions which could occur with the alcohol. What is the temperature of your sand bath? How is your experiment going? When did you start the experiment?

BETTY EXPERIMENT *#2*:

I am not ready to begin serious work in the mineral kingdom in my present lab. I am acquiring equipment on a regular basis, however, and I hope to find a house to purchase before spring. Therefore, I cannot begin the second Betty experiment at this time.

ETHYL ALCOHOL:

I have also enclosed two pages of data on ethyl alcohol. Several items of note: ethyl alcohol cannot be purified to better than 191 proof by simple distillation; 191 proof ethyl alcohol contains a wide variety of impurities, most likely including a small amount of grain sulphur; ethyl alcohol reacts with almost all metals (this needs to be investigated for the second Betty experiment).

SUNLIGHT:

The sun emits a wide range of wavelengths of light, from long infrared to short ultraviolet. The maximum light intensity is in the visible green (about *5000* angstroms). The earth's atmosphere absorbs the extremes. The ozone layer absorbs most of the ultraviolet. The light that penetrates is the visible, near infrared, and a small amount of ultraviolet. Early morning sunlight must pass through more atmosphere and is therefore less abundant in ultraviolet rays than the sunlight reaching the observer at noon. The diagram below shows this effect.

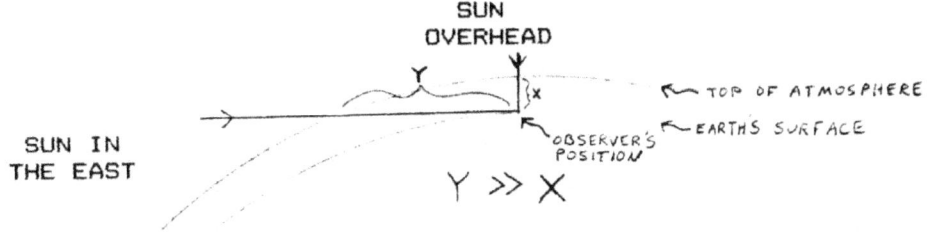

The only ultraviolet light that reaches the earth is in the *3000* to *4000* angstrom range. Exposure of the eyes to ultraviolet light can cause a painful burn of the cornea. Lesser exposure (such as at sunrise) causes less immediate damage but can lead to conjunctivitis (corneal irritation and/or infection; symptoms are itching, smarting, and a foreign body sensation; a scant mucoid secretion might be present; treatment: irritating factors must be eliminated).

Infrared light is also potentially harmful. Excessive infrared can lead to cataract formation. Symptom: progressive, painless loss of vision. When the opacity is in the central lens nucleus, the patient in the early stages may discover that he can see even better than before. Pain occurs when the cataract swells and produces secondary glaucoma.

My point is that staring at the sun is a potentially harmful practice which I advise against.

THE SCIENTIFIC METHOD:
The so-called "scientific method" starts with carefully designed experiments carried out in the laboratory under closely controlled conditions. I do not know, but I am interested in, the laboratory methods and experiments which you and other

alchemical investigators practice. The goal of each experiment should be clearly defined in advance, and a record should be kept in such a manner that the experiment could be repeated with great accuracy by another person. Accurate measurements of mass (weight), volume, and temperature are essential.

It would like to hear more about the experiments performed by you and by your friends. I am willing to share information about my experiments. It is too bad that the Paracelsus publication *Essentia* is no longer issued, as it had a great potential for sharing experimental data. Perhaps we should consider the ramifications of starting an informal newsletter for serious alchemical investigators.

P. N. Wheeler

Reference:

R. Epstein, K. Kustin, P. De Kepper, and M. Orban: *Oscillating Chemical Reactions*, Scientific American, March 1983, 112-123.

"Once thought to violate natural law, reactions where concentrations of key substances periodically rise and fall have now been designed deliberately. They may illuminate similar behavior in living systems."

Ethyl Alcohol (ethanol)

Chemical Formula: C_2H_6O

Boiling point: 78.5 degrees Centigrade (for anhydrous)

Clear, colorless, *very* mobile, flammable liquid; pleasant odor; burning taste. Absorbs water rapidly from air. Miscible with water, ether, acetone and with many other organic liquids. When *ignited,* ethyl alcohol burns in air with a pale blue, transparent flame, producing water and carbon dioxide.

Human toxicity: nausea, vomiting, flushing, mental excitement or depression, drowsiness, impaired perception, incoordination, stupor, coma, death may occur.

Therapeutic category: topical anti—infective; peripheral vasodilator pharmaceutical aid.

Pathogenesis: In general, a linear correlation exists between the severity of liver damage and the intensity of ethanol abuse. Ethanol has a toxic effect on the gut and pancreas. Ethanol is also a hepatotoxin whose metabolism creates profound derangements of the liver cell. (See *The Merck Manual* for further details.)

Most important method of preparation: fermentation
of glucose and related substances. Ethyl alcohol is
produced by the enzymes secreted by the yeast. The
enzymes (zymase) act as catalysts for the breakdown
of sugar molecules. The fermented liquid contains
about 10% to 15% ethyl alcohol which may be
concentrated by careful fractional distillation. The
first distillate to come over always contains
acetaldehyde, an intermediate product also formed by
the yeast. (Acetaldehyde causes a general narcotic
action with symptoms similar to chronic alcoholism.
Large doses may cause death. Acetaldehyde has a
boiling point of *20* degrees centigrade. To remove
the aldehyde, add a small amount of potassium
hydroxide and zinc dust or sodium amalgam, reflux,
then redistill. The result, however, might not be
safe for consumption.)

The maximum obtainable concentration by distillation
is an azeotropic mixture with water containing 95.6%
alcohol by weight and boiling at 78.2 degrees
centigrade. This corresponds to 191 proof.

Other impurities present in the concentrated ethanol
result from protein impurities in the starting
material. These give rise to fusel oil containing
n—propyl, isobutyl, isoamyl and active amyl alcohols
together with smaller amounts of a very complex

mixture of higher alcohols and oily compounds. Fusel oil is about 3 to 11 parts per 1000 of ethanol produced. The residue from the fermentation mixture contains glycerol and potassium salts. If burned, the ash contains about 30% potassium oxide.

To obtain absolute ethanol, the distillate may be heated with fresh quicklime (calcium oxide, CaO) which reacts with water and not with ethyl alcohol, then redistilled. Another method uses a fused mixture of anhydrous potassium and sodium acetates. A method using benzene is commonly used to prepare reagent grade ethyl alcohol. Ethanol of 99.5% can be obtained by a special use of calcium chloride. (The ethyl alcohol concentrated by these methods might not be safe for consumption.) The density of absolute alcohol at 20 degrees centigrade is 0.789 while that of 96% is 0.801.

Industrial grade ethyl alcohol is prepared by the catalyzed reaction of water with ethylene (an addition reaction). This is direct hydration of ethylene. Indirect hydration of ethylene is also used to produce ethyl alcohol.

The oxygen end of the ethyl alcohol molecule is more negative than the hydrocarbon end. Because of their polar nature, liquid alcohols can dissolve ionic solutes. More important is their ability to dissolve molecular solutes.

Ethanol reacts with most metals. It acts as a weak acid, reacting with active metals to form the alkoxide ion and hydrogen gas. Ethyl alcohol is very commonly used in solution for carrying out organic reactions.

A few ethyl alcohol reactions:

(1) With sodium metal, forming sodium ethoxide plus hydrogen gas,

(2) With phosphorus chloride, bromide, iodide, forming ethyl chloride, bromide, iodide, respectively,

(3) With concentrated sulfuric acid, forming at 100 degrees C. ethyl hydrogen sulfate, at 140 degrees diethyl ether, at 200 degrees ethylene,

(4) With organic acids, warmed in the presence of sulfuric acid, forming esters, e.g., ethyl acetate, ethyl benzoate,

(5) With magnesium methyl iodide in anhydrous ether (Grignard's solution), forming methane as in the case of primary alcohols,

(6) with calcium chloride to form a solid addition compound which is decomposed by water,

(7) With oxygen, using sodium dichromate solution and sulfuric acid to form acetaldehyde and acetic

acid; using air, in the presence c-f acetic
bacteria, to form vinegar (dilute acetic acid),
(8) With concentrated nitric acid (free from
nitrogen tetroxide) to form ethyl nitrate; with
dilute nitric acid to form glycolic acid; with
concentrated nitric acid containing nitrogen
tetroxide (fuming nitric acid) to cause an explosive
reaction,
(9) with chlorine (or bromine) to form a chloral (or
bromal) compound.

References:

The Merck Index, Ninth Edition
The Merck Manual, Fourteenth Edition
Organic Chemistry, Frank C. Whitmore
Chemistry, Michell J. Sienko and Robert A. Plane
Van Nostrand's *Scientific Encyclopedia,* Fifth
     Edition

Letter from Manfred Junius, Jan. 18, 1985

SOUTH AUSTRALIA

Dear Hans,

It was really nice of you to send me the interesting pages of information regarding the work of Betty McKaig in California. As always in such matters, it is the practical result which counts. You have seen the result of Betty's work yourself, and seem very impressed with the pulsating alcohol. So are Jim and Art, who went there. I shall gladly try to join all of you in the experiment, as far as my free time permits.

In any case Betty seems to be a most unusual and intuitive person. What you write about her discoveries regarding Philalethes and Bacon is very fascinating, and if, as a former journalist, she was able to arrive at the preparation of such high magistries, we may have a case of intuitive remembrance of her work in her past life close to Bacon. A number of rather extraordinary coincidences seem to come to the surface here.

From the fact that she is willing to share her knowledge with others so freely, it is evident that she is a highly involved person and certainly not a selfish individual. She also mentions a few decisive pitfalls. As you know, you can make the yellow glass of Antimony also with the addition of Borax, as it is taught in Frater Albertus' groups, but Basilius himself mentions further down the preparation without any additives. Moreover, many students of the Paracelsus Research Society extract the Borax afterwards by Soxhlet circulation, which is hot. I have made Antimony tincture also from yellow glass, which I obtained with Borax. Otherwise you need rather high temperatures.

The best is, if we all carry out the necessary experiments. I would love to meet Betty one day, but I do not know when I shall be in America next time. Actually I am only back from Europe since a few weeks, where I had the most wonderful seminars again. We have worked a lot with *Circulata*, and also made chromatographic and capillardynamolitic experiments with extraordinary results. I brought the plates with me and also made slide Photographs. When these are ready I shall send you a picture. In the most beautiful pharmacy of Frankfurt we compared the results of the two Gold Tinctures, one DHA and one my own with a Circulatum. The difference in the capillardynamolitic image is striking.

In one of the seminars we also combined exercises, meditations and above all classical Raga Music with the work. As you know from my books, Ragas, like all true modal music, are musical alchemy.

I shall go again, when I have the time, for further seminars, which I also accompanied by some concert broadcasts etc. All in all a wonderfully stimulating period. You remember I proposed to you once before to introduce you to a little to the alchemy of Ragas.

We also worked with the Circulatum Minus, obtaining various kinds of separations with it from different species. One of the participants proposed the attempt to separate human blood. But as Urbigerus himself mentions this is not possible, since the Circulatum is specified for the vegetable kingdom. But in this way we could clarify the facts.

We had quite a few very advanced alchemists among us, some from Vienna as well. Most of them, in one way or the other, seemed to have learned from Frater Albertus.

I also met my own teacher Pancaldi in Munich, where he had come from Switzerland. As usual, he was very generous with knowledge, had brought copies of Le Breton (Le tres bon) for me together with a lot of advice. For Christmas he sent me a card with the

Green Lion and a message to work on the salts, specially the fixed salts. Since your texts contain advice regarding the Green Lion, I take all this as a suggestion to prepare it in the near future. Now as far as the **Tractatus de Quinta Essentiae Vini** is concerned, Leone has translated it as well. You can find it in a re-print in German called "Schatzkammer der Alchemie." It is a collection of my various texts and has been published by Akademische Druck und Verlagsanstalt, Graz, Austria, 1976, from where you can get it. But the German version of my book on plant alchemy contains it in facsimile print with commentary. It is good to have it besides the work of Hollandus, since there are slight differences in the technique. Please inform Kjell accordingly.

All the best to you, dear Hans, and warmest regards to all brothers and sisters in the Art.

Yours sincerely,
Manfred Junius

Letter from Manfred Junius, Jan. 23, 1985

South Australia

Dear Hans,

While going through my *notes* regarding the making of Antimony Glass, I came across the passage by Schroeder in his **Artzneyschatz**, which I am including in this letter.

The note mentions a special constellation of Sun and Moon as favorable for the work, Aquarius and Pisces. This constellation, in fact, is of great importance in oriental astrology. It is the 26th Moon House, known in Arabic as *Al Fargh Al Mukdim*. It extends from 21° 25′ 45″ of Aquarius till 4° 17′ 10″ of Pisces. I wrote about it in my book, German version page 135.

Besides it is good to start during the New Moon.

Now, you will get this constellation during 1985 only once, and that is on the 20th of February, in the U.S. you will be a day behind. I am including a horoscope for Adelaide, you can easily calculate the corresponding time for your place. Because of the uniqueness of the situation next month I thought I should let you know this fact immediately, so that you have time to prepare yourself for the work, if you want to start that day. Eventually you could get in contact with Betty and tell her as well, and also all our other friends who might be interested to work on the magistery. I shall certainly start making the glass on that day, as well as the Green Lion. It is curious that one of my teachers, Augusto Pancaldi in Switzerland, sent me a Christmas Card with the Green Lion.

This may be a direct suggestion to make some of it.

As far as the questions regarding Betty's work are concerned, I have done some thinking. The idea of a non-violent destruction of the atom with the nuclei floating on top as the Eye of Horus would not be acceptable to modern physics or chemistry, since their material manifestation can only occur through complete atoms, otherwise you would have an explosion. The nucleus by itself would be out of balance. In Ayurveda we would say, that there is a lack of the other mahabhutas.

On the other hand we know that biological transmutations do occur in nature. You may be acquainted with the work of Kervran in this respect.

(Biological transmutations). Necessarily these transmutations would take place at the nuclear level. However, during the process, electrons attach themselves to other partner-nuclei. If this does not take place, there is violence. The hydrogen bomb is based on that type of violence.

However, as so often, our attitude should remain open, I find this the only truly scientific attitude. We still know very little about the true nature of matter, and first of all the experiments should be carried out.

What I would be mostly interested in is the medicinal value of the AA preparation. Remember Basilius suggestions about utility. If we can produce something of use to suffering humanity, that is worth the effort. About the identity of Philalethes a lot has been speculated. Actually the name is supposed to have been used by a number of people, mainly from the 17th to the 19th century. It was also a pseudonym of Helmont. An order of "Divine Order of the Philalethes" was founded 1773 by Court de Gebelin, Savalette de Langes and others. Compare the writings by Frick, Die Philalethes Handschriften des Musaeum Hermeticum (Graz, Austria). The question regarding authorship is extremely difficult to clarify, and Betty must be having her own reasons after all her research. Whoever may be the author, the writings seem to be of great value, and I think you should organize a translation of all of them; perhaps Betty may be of help to you. During my recent seminars in Germany, with many pharmacologists, we had some very interesting discussions. We had been working with the Circulatum, and scientists held the opinion that its working could not be explained in terms of modern chemistry either because of a double abnormality:

1. the splitting of an oil by an alcoholate (normally this would dissolve oils as in ordinary tinctures),
2. the absorption of chlorophyll by the oil split off. (In normal tinctures the chlorophyll goes into the tincture, this is why all alcoholic tinctures from green plants are green).

In case of the CM however, the green colour is in the oil (emerald green). Moreover the oil is a fatty substance, this seems absurd. A chemist from the IBM is of the opinion, that the whole chlorophyll-ring probably collapses during the process, and via a kind of appendix of the molecule the connection with the lipid substance is established. Many of the persons present almost freaked when they saw the action.

There was also a theory about the Ayurvedic
magistery of mercury, which is non-poisonous and of
very effective catalytic action (yogavaha). Dr.Stege-
mann spoke to me at length about the theory about a
pharmacologically inert mercury, which has not yet
been produced in the West, but that a theory exists
about it. It involves a threefold binding of mercury
with itself. Mathematical operations led to this
conclusion, laboratory still has to verify the
ideas. Actually a whole polymer forms in this way,

like a carpet, with bindings like this: $\overset{Hg \equiv Hg \equiv Hg}{Hg \equiv Hg \equiv Hg}$ etc.

We had absolutely fascinating seminars in Europe,
some along with Indian Music which is an alchemy of
sound, as all true modal music. One day I hope to
show you these things.

Letter from Hans, January 28, 1985

Richardson, Texas

Dear Brothers and Sisters:

I am enclosing herein several pieces of data: Copies
of letters from G. "Bears" Hamilton, Phil Wheeler,
Jim Woolsey, Manfred Junius and Kjell Hellesøe.
Kjell has been especially busy translating some
documents. One on the Quintessence of Wine which
will actually be in Manfred Junius' book coming out
this year from Destiny Books. He also translated a
Baconian document, he and I forgetting that this
document appeared In the R.A.M.S. Bacstrom
materials. Sigh. I urge you all to READ that
document. If you do not have the R.A.M.S Bacstrom
"Compendium" (where it appears twice, once by Bacon
and once by Paracelsus!!!)I will send you a copy.
Read it. Reflect VERY carefully on what the
author(s) have to say about the sulfur. And WHAT
sulfur do they speak of? To me, they mean the sulfur
that makes antimony ore stibium trisulfide. But, on
a contradictory note, they manufacture an "aqua
regia" or a corrosive which does, in my opinion, a
"wet" sublimation and drives off the sulfur via
sulfur dioxide. (Ever pour hydrochloric acid on
antimony ore? PHEW!!) Bye bye sulfur!

Now, this is an important document. It relates to the AA of Betty. READ it. What we want is your opinion on how to make the oil. That is, "digest" the process given and put it in your own words. As given, it is open to erroneous interpretation. Since it is so valuable a medicine we want to focus on it and simplify the process and make it clear. So, send me your thoughts. I will coagulate, digest and rectify them!

Dr. Schein, now In Southern Calif. tells me he will visit Betty. Hurrah! He is not only a fine alchemist, but a medical doctor. He already told me a piece of interesting news. That is, the process of the alcohol solidifying (as in Betty's pulsing alcohol) is a known phenomenon. It has a name which he is going to get for me. It has to do with polymerization. Anyway, another little chunk of useful information for us.

When you get this, realize I am expecting some feedback. Pro or con. I want you to consider the article on the Tincture of Antimony by Bacon and Paracelsus. And the Quintessence of Wine translated by Kjell. He has done yeoman duty in this "cause," performing much translation and research: A true alchemical researcher. Now YOU send me your ideas on the Tincture? How YOU think it is made and why the apparent contradiction in the bit about driving off the sulfur. Recall in Prima, when WE made the glass, we calcined the ore and drove OFF the sulfur. Was this an error? The article by Bacon says something to the effect, "the words are true but in practice...it isn't so." What does that MEAN? Your opinions please (isn't this fun!)

All for now. Look forward to hearing from you. Peace.

                        Hans W. Nintzel

Short Note from Hans, Feb. 10, 1985

I got the alcohol to PULSE!

                    Hans W. Nintzel

Letter from Arthur Fehres, Feb. 10, 1985

Dear Hans and other friends of the Art,

Thank you very much for sending me the latest information, as you put it, 'to bring me up—to—date.' As a middle man, your place in heaven seems to be assured!

I am sure, we all realize that if those who received all this material were to express their thoughts and give their comments, your 'job' would soon got out of hand. I therefore propose, that we confine our contributions <u>as much as possible</u>, to that which is factual and what has been personally experienced, either within or in one's lab. If not, we could be flooded with reading material, which eventually will defeat the very purpose of our valued communication.

Timewise it is impossible for me to go into your questions regarding the said Paracelsus/Bacon tracts to any great length. However, what I do write about this, should be considered <u>worthwhile reading</u> as I dare say: I have personally produced the volatilized version of 'AA,' but before I go into your questions, though, I feel I should clear the air, about what was written regarding homeopathy, language of the Gods and myself.

As a natural healer with diplomas in naturopathy, chiropractics and Chinese acupuncture, I also practiced homeopathy extensively for 15 years with remarkable results. Obviously, whoever commented on what I wrote about homeopathy, (I could not read his signature) has gained no worthwhile practical experience in this field.

It is also obvious, he has not come up with the volatilized oil of antimony, or he would not have written about homeopathy as such. One drop of the alcohol, gently distilled of THIS oily tincture, sends one 'sky high' and one receives so much energy and alertness, that one feels no need for sleep! At least, that following night. It has to be triturated because of its tremendous strength (=potency). He admits himself that sooner or later one may have to resort to techniques similar to those in homeopathy after perfection of a preparation. However, to evaluate its 'perfection,' involves the trial and error principle, which can now be minimized, through the THERATEST. I promise more information on this as it comes to hand.

Secondly, whoever rubbishes language of the Gods or phonetic cabala, rubbishes Fulcanelli as a Master of The Art. Calling it 'mental masturbation' clearly

shows the stage he is at. So far he has indeed only
fooled around with it, like children playing with
words and cracking little jokes. Do not think I am
offended. Not in the least!

Regarding the said Paracelsus/Bacon tracts I would
first like to point out, that it does not really
matter who the authors are. What is more important
is what is written and how it is put into words.
When one looks into THIS deeply, one will find it of
great help to separate truth from grudging. It
applies to all tracts. Reading a tract requires
great caution, oven more so when it is a
translation, and most of them are.

As an example let us compare the first part of the
two corresponding sentences:

1. Paracelsus tract p.7 'The meaning of these words
is good, but...'
2. Roger Bacon tract p.5, 'These words and opinions
are good and true, but...'

Re No. 1: This is correct, because the meaning of
all words is good. Any word has a meaning which is
good. Good means 'correct.'

Re No. 2: This is incorrect, because of the words

'opinions' and 'true.' An opinion is either true or not true. 'These words' refers to the previous paragraphs, not to single words, so 'these words' can also be true or not true.

Secondly, to me, this part is unmistakable proof, that Bacon's tract is also a translation and that the said part is <u>a mistranslation</u> either from Dutch or German, in which we find the expression 'goed en wel' and 'gut und wohl' respectively. This expression is used when one disagrees or at least has some objection and <u>out of politeness</u> they first agree, but in fact they disagree. The disagreement directly follows the expression which starts with 'maar and 'aber' respectively, corresponding with the English 'but.' Personally, I would reword the said part: "They mean well, but..."

Let me now come to some of Hans' questions:

Q1: "The article by Bacon says something to the effect, 'the words are true but in practice… it isn't so.' What does that MEAN?"
<u>Answer</u>: In both tracts it means that the author does not agree with vitrification before volatilization, 'For glass is the end and last of all things' (tract *Paracelsus)* and 'for glass is in all things the utmost and last' (tract Bacon).

Q.2: "What sulphur do they speak of?"
Answer: They speak of the most volatile part of the solar sulphur.

Q.3: "Recall in Prima, when WE made the glass, we calcined the ore.

    (3/a) and drove OFF the sulphur.

    (3/b) Was this an error?"

Answer:

3/a: No.

3/b: Impossible to answer because of the wording of the whole question. The answer would consist of many ifs and buts and assumptions as to what Hans means. For instance, I can assume that with 'the sulphur' he means S_3, because he states "to me, they mean the sulphur that makes antimony ore stibium trisulphide."

Secondly, does he really mean (as the wording of the question implies), that he truly calcined the ore, before making glass?

Thirdly, I don't know how HE made the glass in HIS Prima. Albert showed several methods.

As I have stated, I made the oil AND simplified the process as well, but as it in an ART and a SCIENCE.

It requires oral instruction. Besides, to write it all down as precisely as possible, is far from a guarantee you will come up with it.

I can see, therefore, not much use in further correspondence as far as THIS subject is concerned.

Fraternally yours,

Arthur George Fehres

Hans wrote the following answers to Arthur's Questions

Yes, I meant Sulphur (S) = 'Brimstone.' As per George Starkey and others this sulphur is highly medicinal. Could the 'Philosophical Sulphur' (♁) be driven off by an acid? As to the glass at PRS, we took ☿ ore, put some in an earthenware dish over a Fischer burner. It was stirred to break up globules that formed. Sulphur fumes came off. It turned to a tan-gray-white color. SbO_3. We took this, put it in a crucible, fired it, it melted; poured it out as yellow glass. Anyone out there make ⚭[8] of ☿ yet?

---

[8] Oil. -pnw

Letter from Kjell, 1985

Jim has a theory that physical nature is merely a social phenomenon! Well, that may be a workable theory if you are the leader of a sect, with a small "society" to control, but otherwise who benefits from such ideas? Such ideas are certainly very few & far away from the "Weltbild" to be found in alchemy. But he later on agrees himself that are great if you "spin your wheels" on the mental plane, but fails to materialize positive results if you wish to set the alchemical millstones grinding out physical results. (What is GL anyway? = Green Lion?).

Bears says that the geomagnetic field may have an oscillatory frequency (40 Hz) and thus be "driving" the oscillatory phenomenon, because it is "aligned" along the N-S axis.

But although the alignment may very well be a magnetic phenomenon, this does not necessarily mean that this is also the driving force. No, I think the driving force is much more obvious that that; it is simply the heat coming from the sand bath. Slowly complexes may be built up under action of the heat, until a certain degree of saturation is reached. At this threshold some kind of chain reaction is set off which drastically reduces the density of the molecules built up in the previous "up" phase. This massive disintegration goes with the emission of light. Then the supply of built up molecules being ended, the "down" phase is over and the build-up starts again under the influence of heat; and there you have your cycle, flashing on-off, on-off, etc.

Electrons have certainly got spin (as discovered by Goudsmit & Whlenbeck 1925[9] and therewith associated with a <u>magnetic moment</u>. So electrons actually are small magnets. Now if you have an <u>even</u> number of electrons, they may be paired

so that the magnetism <u>cancels out</u>. But with an odd number of electrons you may actually get an atom or molecule to be a small magnet. But whether the idea of electron spin will prove to be useful in alchemy in the near term I doubt, unless you actually start to experiment by actually putting your tinctures (on their hot plates) in <u>very strong</u> magnetic fields. (Buy a big magnet that weighs 20 pounds!)

But electron spins <u>may</u> hold the key to the mechanism of alchemical (i.e. low temperature) transmutation. Because the spins will act as "doorkeepers" to let exchanges propagate between adjacent nuclei, separated by the otherwise impenetrable (at the energy levels in question)

[9] Ralph Kronig working as an assistant to Alfred Landé, according to Wikipedia. -pnw

electron shells.

This last sentence contains some very important ideas, and should be communicated to Junius, because his ideas on transmutation are entirely in accord with mine. He obviously has given this a great deal of thought. He is unusual in that he both has a grasp of alchemy, as well as the principles of modern science. Such people were hardly to be found during my time at the P.R.S.[10] at least as witnessed by the ones I actually met or those who ventured to publish in **Essentia** or the **Bulletin**. Usually they were "Spinner (German)" when it came to modern science in relation to alchemy.

With regard to the tract "de Oleo etc…" I'll have to do some more thinking first.

But obviously an acid has the same <u>oxidizing</u> effect (i.e. it increases the metallic oxidation number as the chemist say), as you also say.

Quicksilver being the "Mother of the metals" is obviously wrong. But you must certainly know this after all these years. You are only saying this because it gives you pleasure to hear what kind of argument I can come up with. That Quicksilver dissolves copper is no miracle, but even explainable in terms of ordinary, commonplace chemical potentials, as I pointed out to you before. However it <u>is</u> true that some alchemists of some renown <u>used</u>

---

[10] Paracelsus Research Society. -pnw

Hg in some of their key processes. They may even have used it as a "carrier substance" for some other, very important transmutatory reactions. In other words they <u>mixed</u> it with some other partly <u>secret</u> ingredients and thus achieved their goal. Dr. Joachim Voigt also told me that he had witnessed a transmutation by a German Alchemist after the war. He is a member of the Saturnian Brotherhood, and the AA I think. He had crossed the "abyss" (Binah) and wore a black ring with a gold serpent as a visible sign of his attainment.

Mercury is the Mother re-absorbing her children, you say. In classical mythology this refers to <u>Chronos</u> who <u>ate</u> his own children. But Chronos was <u>male</u> and <u>not</u> Hermes (who was the son of Zeus and Maia). Recall that Zeus (= son of Chronos) was <u>saved</u>, because Chronos was <u>fooled</u> (like Esau by Jacob) and given a <u>stone</u> to eat. Later on the stone was <u>spit out</u> and landed on top of a mountain (HELICON).

That alcohol "polymerizes" is certainly not new. I asked a professor at MIT about this in 1974. He said that there are many bigger, larger molecules than the ordinary C_2H_5OH (ethanol), but they are usually not mentioned in introductory undergraduate chemistry texts. He said that whiskey (good whiskey) is so famous exactly because it contains these beneficient "higher" alcohols (i.e. larger molecules) that are volatile. Therefore the art of distillation is so important in the manufacture of high quality liquors (<u>and</u> its digestion); Among other things, <u>copper</u> stills must invariably be used. Apparently the copper influences the fermentation and maintenance of these higher alcohols.

I pointed out to Frater Albertus in 1975 that the antimony glass acted as a <u>catalyst</u> in forming these higher organic chains on the basis of ethanol. I also told this to a whole bunch of people at the P.R.S. both in America and in Europe. So this fact ought to be well known by now.

Also, it is very well known that alcohol forms "complexes" with metals. This should account for many of the so called "tinctures."

Consider my own experiment where I obtained a KERMES-coloured <u>precipitate</u> from alcohol with alkalis, which I described in my long article I sent you in December (at the end thereof). Now ask someone what <u>that</u> means. (Probably a simple explanation.)

Letter from Kjell, Feb. 17, 1985

As I sit here and write, I realize that one of the very secret alkahests may have been under my very nose for many years now!

Only my lack of curiosity in daily phenomena prevented me from following out this line of thought.

But my inquisitiveness knew no limits, and this finally resulted in a telephone call to Paris where I think an important clue was found.

But I'll give you more details in a later letter.

I have not heard much from you lately (compared to all that's been happening spagyric-wise), but I guess you're rather large correspondence plus lab work, all beside your regular job, keeps you quite busy! All in all a very admirable effort.

By the way, did I send you page 1 of, I think it was Pancaldi's German notes on tartar, antimony and some other things? Back in 1977 I was working on translating these (you were even there; I started it in the Quarta class!). Apparently they were finished sometime soon after that, because I just found the complete translation among some old papers! Well, I think I sent you the first page last month.

Hans, do you think you can find out whether Frater Albertus has a curriculum for his seven classes? I think he had one, but from some remarks in his book, "Alchemy in the 20th Century" plus remarks by other students I have met, it seems as if he did not follow a curriculum <u>rigorously</u>. Perhaps you can collect class notes from early students, and try to reconstruct the original full curriculum? Don't you agree this would be a worthwhile project? E.g. what do we know of the "vinegar of antimony" (see Pancaldi's notes if you already have them). But this is all terribly vague. Now in the chemistry books there <u>is</u> an acid of antimony. But is this the one that Frater and Pancaldi and Basilius hint at?

What people that you know are worthwhile to ask such questions?
Kurt?
Dale? (Do you know what he is doing?)
Junius? (I sent him a copy of my QEV tract.)
Dan? (He was not exactly a sponge that you could squeeze when I met him in 1975.)

(There must exist people who have answers.)

Oh with respect to Arthur's "sun-gazing." Do you know that Carlos Castaneda recommends the same? (but with left eye only). You might ask him this next time you write him.

I have another translation coming up. I hope you are interested in this one too! But I sorely need a good proofreader. What did you think of the quality of the QEV tract? Did you read all of it yet?

Articles on the identity of Eiraneus Philalethes have been published in Ambix:

Vol. XII, No. 1, p. 24 (1964): The problem of the identity of Eiraneus Philalethes. (See also Vol. XIII p. 52).

Vol. XIX, No. 19, pg. 204 (1972): Further Thoughts on the Identity of E.P.

Perhaps, if you have access to Ambix, which I have not, you could find out the address of Mr. Wilkinson, who is an American social and literary historian, and obtain copies of his articles and possibly a personal account of his present opinions.

Personally, I always visualized Philalethes as a Dutchman, because his first work was printed in Amsterdam (in 1667). In fact I always visualized him as the man who so mysteriously visited Dr. Schweizer at the Hague on December 27, 1666. (See the Hermetic Museum, Helvetius' Vitulus Aureus or Golden Calf). But this is only the way I would like to see the truth unveil, <u>if it was up to me</u>.

Dan's effort to bring out a new translation of the Introitus Apertus is admirable. I personally have versions in both German and French, beside the original Latin version. If I have time, someday I will compare them. I feel sure that at least one of them is reliable. After all, many people were conscientious, even back then.

With respect to your letter from Dan on the possible identity of Philalethes, I have some remarks and references.

First of all, if it could be established beyond reasonable doubt that Philalethes was John Winthrop Jr., this would amount to a fantastic breakthrough. At the same time I feel that many of those involved in alchemy would not be willing to go along with this idea, since it would somehow make him less mysterious, and the image of the mysterious adept is something that many would not be willing to let go of.

Dan's assumption that Philalethes was not Francis Bacon, is something I should like to support. I see that there is a tendency in some quarters to try to identify famous but unknown authors and artists with people from the upper levels of society; such as those who claim that F. Bacon was Shakespeare etc. But is this not like claiming for instance that e.g. Jacob Boehme was identical with King Frederic V of Bohemia? But the case of J. Boehme definitely shows that even those belonging to the lower classes at times by far surpassing those who had been brought up at the wealthy courts by the best teachers available and with plenty of leisure time to pursue their literary and experimental interests.

In the case of the identification Philalethes equals J. Winthrop Jr. (1606-1676), I wish that Dan had supplied some more details as to how he was able to identify Philalethes' own handwriting. After all, before the tracts were printed, and the handwritten manuscripts were circulating, the only way to copy them was by hand (they had no Xerox machines!). My point is that you came across a handwritten version of one of Philalethes' tracts, how can you know it is the original? My not have J. Winthrop Jr. have borrowed one of the circulating manuscripts from a fellow alchemist and hand-copied it himself for his own use?

Kjell

Notes from Kjell

I found some notes on a piece of paper that were
related to my letter to you in January. Did I
already send this to you? I'll repeat it in case I
did <u>not</u> already send it.

Spieß   = spear
Rütteln = to shake
der Speißrüttler = Shakespeare
glanz = Lustre
der Igel = Hedgehog
Stachel = spine

Alfred Freund wrote a book titled "das Bild des
Speerhüttlers, die Lösung des Shakespeare rötsels"
(Hamburg 1921). In his book Bacon is the
philosophical Hercules, whom time will reveal is the
true "Spear-Shaker." Freund's book is based on the
following book by Blaise de Vignère (1522-1596):
"Les Images ou Tableaux de platte peinture des deux
Philostrates sophists grecs et les Statues de
Callistrale" (3rd edition Paris 1637). On page 486
occurs the "Hercules Furieux." Blaise de Vignère is
famous for his works "Traité de feu et de sel" and
"Commentaries sur Philostrate." He was a noble from
St. Pourçain Bourbonnois.

Notes on Components of the Philosopher's Stone

Selenite

The name originates from mythical Greek figure Selene, meaning MOON. In myth, Selene crept nightly into a cave where Endymion, a beautiful youth, lay sleeping and there "made love to him." (Love=eros, attraction; implies sane sort of electromagnetic effect on passive element(s)?) Selenite is the moonstone of Pliny; moonstone said to wax and wane (increase and decrease in size or volume) with the phases of the moon.

Selenite is a gypsum type of limestone. Selenite crystallizes in sword—shaped blades that radiate from the matrix. May be infinitely small or giant in size, e.g. two-foot blades from the Cave of Swords in Mexico.

Selenite is a salt of selenius acid, a sulfate of lime; dibasic acid. It partakes of the peculiar properties of selenium: this mineral is <u>a conductor of electricity under light and a resistor in the dark. It polarizes light</u>. Salt of Selenious acid prepared by fusing selenite/nitre. (Note: this component, like the other two, is in <u>sulfate</u> form.)

Selenic acid, soluble, forms metallic selenates, less stable than selenites. "The acid of vitriol (sulfuric acid) forsakes its alkali to unite and form a selenite with calcareous earth. Thus sulfate of lime or selenite may be precipitated with alcohol from water which contains this salt.

Selenite is EX CALIBUR, the magical sword of Arthur, frozen in the stone. The name derives from the Gk. X=chi=fire; Cali=hot; burn=burning: burning hot fire. The alchemists derive Halicali= holy salt.

Of the biblical "Salt, sulfur and mercury," this appears to identify, specify the first. The chemical heat of limestone in selenite form constitutes part of the hidden furnace of the alchemists, i.e., the "triple fire." All indications are that it is HELL FIRE, into which the descent is made.

SELENITE is a member of the sulfur family.

Selenite occurs in a second crystal form called "ram's horn," so called because that's what it looks like, i.e., a curving tube form; this is the origin of the Shofar or Rain's Horn which sounds for the Hebrew high holy days...in autumn.

COMPOUND 3

Jerusalem and all of the Holy Land are situated on an immense limestone formation; the New Jerusalem is said (in the Baconian literature) to be in Nova Scotia, which lies on the great Windsor Formation of limestone.

This is a considerable jumble of information, but I make no apologies for it, since it is to some extent undigested matter and to a great extent unorganized material as far as a written account is concerned. However, the time it would take to properly organize for a more coherent account would preclude the continuation and completion of the experimental work, which must take precedence over all.

These notes are presented with the hope that they will be received, not for disputation and argument but to be weighed and considered in the true spirit of Bacon's "Sons of the lamp." To those seriously interested in experimental alchemy, they should be of immense help.

Random notes: Jim Woolsey brought me some of his kermes alcohol at Christmas time and I set some of it to incubate; it has reduced to merely cover the bottom of the flask but remains a pale, watery orange in color instead of the blood—red obtained with absolute alcohol or antimony purified alcohol; there is no indication of pulsing.

Since Jim tells me the kermes is made with lye, it is probable that nothing profitable can come from this substance. Eirenaeus specifically warns against the use of caustics in preparation of the materials. This I suspect is because the alchemical process involves the generation of living fermentaceous entities that evolve the enzymes that run the cyclic wheel of generation and decay.

It should be understood by alchemical students that the literature on the subject is written in code, and as in all good codes, the security of the message is to a large extent dependent upon the use of nulls or nonsense phrases interspersed with the true message. The student is expected to develop sufficient sharpness of wit to spot "the flies in the ointment," where incongruities and red herrings are strewn in the literature.

For example: Where Basil Valentine presents the method for preparing antimony through the yellow glass and the extraction of antimony orange by vinegar, he tells us that this is the very best way he knows, but then goes on to offer a number of obviously <u>inferior</u>, and sometimes obviously <u>foolish</u>, ways of doing the same thing.

Q.   Why would anyone want to approach the process by any but the <u>very best way?</u>

Mythology, the Bible, and Alchemy

ADAM = a fine red dust (antimony orange)

EVE = the breath of life

    Eve taken from Adam's rib; rib/bone= calcium carbonate.

MOSES (in the bullrushes) bullrushes, reeds, grow in marshes which are brackish...the littoral in which fresh water runoff merges with the salt sea. Sulfuric acid, an essential component to the formation of metal sulfates, forms as a slimy matter around reeds in brackish water; the Egyptians harvested it by hanging, knotted strings around the reeds upon which the slime would collect and could be taken by stripping the strings into their collecting vessels.

MOSAIC means made up of many small particles of

stone, *or* colored glass. The following from Van
Nostrand may be applicable:

MOSAIC STRUCTURE: Evidence from X-ray analysis
suggests that even an annealed single crystal is
usually composed of a mosaic of blocks of the order
of 5000 Å in dimension and tilted to one another at
angles of the order of 10 minutes.

Letter from G. S. Hamilton, March 3, 1985

Dear Hans,

I was very excited to hear that you had got the alcohol to pulse! I've been marshalling information to reply to your question concerning the oil of antimony. I have quite a few things to say.

The Paracelsus/Bacon articles are obviously redundant; who knows where the original comes from. It seems clear, however, that this is the same process in Basil's Triumphal Chariot[11], pg. 99-102 (London 1962 edition). Yet another variation is the one given by Frater in an old **Parachemy** under the title Pure Sweet Oil of Antimony. (I forget the issue number but enclosed is a hand copy of it.) There are various differences in these processes and I think they are consistent with one another. Sulfur does not always mean elemental sulfur; it also refers generally to the property of acidity (that is, the property of having a low-lying empty electronic orbital).

Basil uses glass of antimony; presumably, stibnite roasted to the fused oxide (Sb_2O_3 or

---

[11] The Triumphal Chariot of Antimony, Vol. 2, The R.A.M.S. Library of Alchemy. -pnw

SB_4O_6)! He treats (digests) it with vinegar (apparently not vinegar of antimony but acetic acid). He then decants off the golden tincture and distills off the vinegar to obtain a gold-red colored powder; extraction of this with spirit of wine provides the oil.

I suspect the solid compound obtained from the vinegar is some antimonial compound (containing SbO^+ mostly). Such antimonial salts may be obtained by dissolving the oxide in the appropriate acid and crystallizing. For example, tartar emetic [potassium antimonyl tartrate, $K(SbO)C_4H_4O_6$] is obtained this way via tartaric acid.

Note that in Basil's process there is no treatment with mineral acid prior to the digestion with vinegar. Paracelsus/Bacon differ in that they treat stibnite (Sb_2S_3) with HNO_3, precipitate it out, and take this compound and digest with vinegar and proceed as with Basil.

Finally, Frater starts with Sb_2S_3, treats with vinegar of antimony, obtains a solid which is then treated as before (i.e. wine vinegar -> red solid, extract this with ethanol). Do all these differences make sense? Yes, I think so!

One wishes to perform the digestion with acetic

acid with an oxide of antimony; specifically, an oxide of Sb^{+3}, Basil's glass is just such an oxide and he begins directly with it.

Paracelsus/Bacon seem to be warning that obtaining the oxide by the thermal oxidation route (roasting stibnite in air) is fraught with danger. I don't think they are saying that this won't work; but that a little overmuch application of heat and the desirable properties are lost. (I don't know in what way or what change is indicated.) They thus suggest an alternative method to obtain the oxygenated antimony; they dissolve the Sb_2S_3 in mineral acid (HNO_3/HCl, powerful oxidizing solution) and precipitate with alkali (they don't specify which), this will certainly liberate the sulfur (as you noted) and provide the requisite oxygen. So, I feel that it is doubtless that sulfur is removed regardless of the method; an oxide of antimony is obtained. The acid dissolution is just another way of doing it to avoid the pitfalls (whatever they are) of the air-roasting route. When he says "The words are true…but in practice it isn't so" I think he means that it is possible to roast Sb_2S_3 to Sb_2O_3 and use this, but in practice an error is so readily made that this method is advised against.

Another interesting note: they (Paracelsus and Bacon) warn against leaving the Sb in acid solution too long or you will ruin the tincture and have only a "slick, soft yellow mud." If left in solution too long, one may obtain $SbONO_3$, which will not give the tincture with acetic acid (it is the salt of a stronger acid).

Finally, Frater starts with Sb_2S_3, treats with vinegar of antimony, and obtains a compound which he makes into a tincture with acetic acid. In this case the vinegar of antimony (very, very different from acetic acid, wine vinegar) performs the same trick as the HNO_3/HCl. Thought the composition of vinegar of antimony is not yet known, it is certainly an OXO-acid of sulfur; from this treatment one obtains an oxygenated antimony compound, again, a salt of an acid. This compound is then "extracted" with ethanol to obtain an oil.

The end product is the same in every case; one has an antimonial salt that catalyzes the production of an oil from ethanol.

I have a lot of notes on the properties of antimony and its compounds and I'm going to send a brief exposition of this data to Phil. If you want, I'll send the same to you.

I've looked at the German manuscript you sent
me and translated the first bit in a crude form. I
think I'll be able to handle it; the worst are the
recurring Latin phrases!

Well, I'd write more but this is late getting
off already so I'll hold the rest of my thoughts
for the next letter. This addresses the main issues
you asked about. Hope to hear from you soon and
take care!

G. S. Hamilton

Letter from G. S. Hamilton, 1985

Dear Hans,

Sorry to be so long in replying to your last letters! Things piled up in research work before Xmas (reports to write, experimental details to wrap up, etc.) And, of course, the Xmas-New Year's holiday merry-go-round itself can be pretty disruptive, though fun. Since getting back to the lab 1985 has consisted of 12-14 hour days of work. Lots to do on all levels/planes!

Let me tick off a few items:

First, feel free to send me some German pages; I'd love to try my hand at translating them. I've had German as an undergraduate, and have translated modern research papers. I'm ready to tackle some archaic stuff. My fiancée (also a chemistry major) has had German and is in fact of German descent.

Along similar lines, have you heard of a book called <u>Trois Anciens Tracts</u>, compiled by Canseliet? It consists of three fairly short treatises in French. I haven't heard of the tracts in any other place. I'm working on translating them from French; one is almost done, another ~½ finished, and the third only a few pages. Any interest in the finished products? Needless to say, they are available to R.A.M.S. if you are. One is an explanation of cathedral sculpture in light of alchemical symbolism (sort of like Fulcanelli's book). One is called the Terrestrial Heaven, a general treatise on alchemical theory; I forget the title of the third at the moment. Let me know what you think.

I read the Betty story with great interest and have thought about it a great deal. I also read Arthur Fehre's comments. My thoughts are not complete but here are a few:

It is a basic alchemical tenet that new life arises from the ashes of the old; putrefaction (a form of destruction) is the key to all alchemical transformation (as Basil, the Golden Chain, and every other alchemist states).

At any rate, the important fact is that the oscillations run along the magnetic lines of the earth (N-S) and not normal to them (E-W). Presumably a substance with magnetic properties has been created. It appears to resonate with some regular fluctuation in the earth's geo-magnetic field (an oscillation of 40 cycles/minute).

I've more to say on all subjects but if I tarry too long on this letter I'll get sidetracked by other work and another month will pass. I've been meaning to write for weeks. So I'll fire this off now and start collecting thoughts for the next communiqué. Let me know how things are on your end.

G. S. Hamilton

P.S. In my next letter I'll send a few comments concerning the biochemical basis of cancer.

Letter from Daniel

New York

Hello Hans:

The following are a few comments prompted by the
interest aroused by The Betty Story.

First, and with due respect to Betty, I have to take
exception to the statement that Eiraneus Philalethes
and Francis Bacon were one and the same. Not so.

This is a subject very close to my heart. Ever since
I came to the North East I have been not only
studying the works of Philalethes but I have been
profoundly interested in the life of the American
Adept. That is right, American, although he was born
in England, he made America his home and was no
other than John Winthrop, Jr., the first Governor of
the State of Connecticut, and son of John Winthrop
Sr. who was the first Governor of Massachusetts. To
consider him but a 'nom de plume' or another
pseudonym of Francis Bacon does a terrible injustice
to the great soul that was Eiraneus Philalethes.
This I feel should not be perpetuated or encouraged.

Since, then, for the past twelve years or so I have
been researching the life of Philalethes let me

share a few things with you. I must say though that Francis Bacon is another of my great interests, therefore the response that Betty's comment elicited from me.

Let us look at the chronology. Francis Bacon, Lord Verulam, Viscount St. Albans, son of Queen Elizabeth and the Earl of Leicester, died in 1626 and was buried in Lichfield, at the Charterhouse, in Staffordshire, England. Granted that in all probability a mock funeral was enacted, and Bacon had in reality undergone a 'philosophical death,' and then continued with his mission in the Continent.

Philalethes by his own admission had attained the Mastery of the Work in 1645, when he was but 23 years old. His birth, natural not 'philosophical,' is very well documented. The 'Introitus Apertus...' or 'The Open Entrance to the Shut Palace of the King,' his inspired masterpiece was published for the first time in 1667 in Latin and in 1669 in English by William Cooper. By the way this small treatise, more than anything, created the most intense interest in Alchemy ever witnessed in England, for the next 30 or so years more books were published on this subject than in the whole history of the English language. But in any event, more than

twenty years after Bacon's death he was still
working on the rough draft of several of his works.
I have examined his notes, letters, and very much
traced the majority of the books in his library,
many of those published long after Lord Verulam's
funeral and with copious marginal notes exhibiting
the peculiar handwriting of Philalethes and the
symbols be used. This material is now scattered
throughout New England, New York and Pennsylvania,
with some correspondence and notes in England, but
it is accessible nevertheless.

Incidentally, I have examined four manuscript
versions of the 'Introitus,' these were circulated
long before Langius publication and apparently W.
Cooper was aware of this. All four versions differed
somewhat, these were in Latin. The two English
renditions, Cooper's and Arthur Waite's seem to be
quite inadequate. I am presently working on a new
translation of this priceless work, which, I hope,
will be of some value to other students.

Going back to our subject. More important perhaps
than the chronological discrepancy is the vast
difference between the writings of Bacon and
Philalethes. Surely the styles of writing don't even
resemble each other. And the message or the theme is
simply not the same. Philalethes' works are single

minded to the point of obsession, that is, the confection of the Philosophers Stone in the laboratory, nothing else seems to obscure his message which rings loud and clear with the intensity that only inspiration and first-hand knowledge can provide. His works are in perfect harmony and complement each other, what is obscure in one is perfectly clear in another, what he starts in one he finishes in another, and so on.

The works of Bacon are another matter. The significance and the impact on our whole Tradition of the mission of Lord Verulam is, even at this time, difficult to conceive. Suffice to say that we have here one of the emissaries of the Inner Orders, entrusted with the knowledge of the Ancient Wisdom AND WHOSE MISSION WAS THE PERPETUATION AND DISSEMINATION of the Mysteries. As a leading figure of the Rosicrucian Movement all of his writings are permeated with the ideals and principles of the August Fraternity. And, yes, his works are riddled with references to Alchemy, but Alchemy of a different order, not merely the laboratory praxis but far beyond. I am afraid I must stop here, we are venturing too far afield. Maybe another time. But before we leave this subject let me just mention a possible clue as to the source of the Mysteries that Bacon was to preserve and disseminate, and that is the faithful meeting of Lord Verulam and Giordano Bruno. This defrocked monk bad presumably been able to visualize the Egyptian Mysteries for what they were, and was eagerly trying to communicate the meaning of Hermeticism in terms of the Mysteries of Egypt, it cost him his life. Now if we observe the Rosicrucian motto EX UNO OMNIA, and the word OMNIA has among other senses the anagram of the Spanish word I-AMON, defined in 16th Century Spanish as 'a gammon of bacon,' I-AMON. Thus Amon and Amon-Ra may be valuable subjects of inquiry in connection with Bacon. But even with the word OMNIA alone we could probably spend a great deal of time, it may be a key to the Rosicrucian symbolism of rebirth. But enough.

Going back to Philalethes. No, he was not George Starkey, not Thomas Vaughan (Eugenius Philalethes), nor the noble author of 'The New Atlantis.' But Dr. John Winthrop Jr.

If anyone is inclined I would appreciate hearing on the subjects of Philalethes and Lord Bacon.

I would like also to comment briefly on Arthur Fehres interesting contribution.

Regarding the 'AA.' Being familiar with Homeopathy myself, I realized, like Arthur, that we are dealing with a homeopathic substance, plain and simple. Now, the question is this, do we pursue this substance in terms of Homeopathy? To what extent would this be worthwhile?

Let us digress. A homeopathic substance when used as a remedy affects very few people in a positive or healing manner, the majority of people experience no effect and finally some may experience a drastically adverse reaction. It all has to do with the constitution of the individual (homeopathically speaking), his/her tendencies or predispositions. A homeopathic remedy is prescribed in terms of a complex set of symptoms. It is for this set of symptoms that the remedy is "proven," that is a healthy individual is administered the substance in excess, until some definite symptoms arise, which are, I must add, extremely specific, for instance it may be that a head ache results on the left side of the head, nausea, aversion to some foods, etc., etc. A patient with those symptoms, as exactly matched as possible, would be conceivably cured by that remedy, it then constitutes the "similum" from 'simila similibus curantur' or 'like cures like,' which is the principle of Homeopathy.

There are more than three thousand 'proven' homepathic remedies! The 'AA' would have to be

proven to be of any value. Each of the homeopathic
remedies in the existing repertory can be enormously
potentised. Hahnemann, with a stroke of genius,
taught how to release the healing quality of any
substance. By the way, he insisted that mere
dilution is not sufficient to potentise a substance,
it must be subject to 'succussion,' a violent form
of shaking or percussion of sorts. In any event to
end up with just another homeopathic remedy seems
hardly worthwhile.

Bottom line. In my opinion the 'AA' falls
considerably short of the mark. If you have come
this far, push on. Let us evaluate the spagyric
preparations as such first, and we have a long ways
to go here, what I mean is let us have a perfectly
and completely prepared substance and see what it
does. Sooner or later we may have to resort to
'proving' techniques similar to those in Homeopathy,
but for now we should perhaps concentrate in
perfecting the preparations.

There is much more I would like to add but enough
for now.

Daniel Tracey

Letter from Hans, March 13, 1985

Richardson, Texas

Dear Brothers and Sisters:

Some of you may know and remember Rick Stern from P.R.S. He called me the other night. He had moved to Sedona Arizona in order to be near Dr. Francis Israel Regardie. Doc, you may recall, was a noted author and VERY interested in alchemy. His latest literary effort, "The Complete Golden Dawn System of Magic" included an article on alchemy that I wrote. Rick and Doc had gone to dinner at a friend's house and during the time, right after a gourmet meal, Doc passed through transition.

All those who knew Doc are very sad, very disappointed. For me, I had lived on his street in Los Angeles and we became quite close even though we now lived separated by many miles. I looked upon him as a confidant, an expert who could answer my questions, a friend. He is gone, but he left us in a wonderful way. With friends, enjoying a meal, peacefully. His heart failed due, they believe, in part to the lung damage he sustained doing certain alchemical experiments. He was 78 years old. A triple Scorpio. We will miss you, Doc.

This has been a period, the last twelve months, of many known and noted "occult" figures to pass: Baba Muktenanda, Frater Albertus, Jane Roberts, Gopi Krishna, my friend Dave Kennedy, Doc and a few other psychics and so on. Interesting as to what this might portend.

Back to the present, enclosed are copies of letters, etc. from our friends who are involved in the study of the "pulsing alcohol" and "antimony purified alcohol." I had hoped to get a long letter from Betty but she had hurt her back and was not able to sit and type. Next mailing, we will have quite an update from her.

In the meanwhile, do send me any ideas, suggestions, discoveries, etc. on this subject matter. Between us we will either come up with some valuable data or lay the matter to rest. Thanks for all the support thus far. I might point out I have been taking the AA for two months, putting a few drops twice a day under my tongue, and I have not had any noticeable changes. (other than I became pregnant. Just kidding!)

To fill out this page, I just acquired "Theatrum Chemicum" a six volume set from the mid-century time

frame that has hundreds of old alchemical tracts.
Yum. Too bad they are all in Latin! I am going to
see if I can have certain "goodies" translated. If
you are aware of any of these that are especially
"good" or "useful," I am open to ideas.

Hans W. Nintzel

Letter from Hans, April 3, 1985

Greetings Brothers and Sisters:

I am enclosing, for your review, some papers prepared by Betty. These are not all of what she sent but it will give you the "flavor" and convey the main information to you. What I did not send was notes on Antimony, sort of technical notes from some reference, no doubt "flavored" with her own personal observations. If you are eager to have these, let me know and I can *Xerox* them for you. However, while interesting, you have probably got this data already and I am reluctant to invest in copying it all.

I am trying to "connect" with a Mary Lynn who lives in Ohio and whom, I am given to believe, has done this "AA" work following the advice of Fulcanelli in one of his books! I have written her and will see if she has indeed done this work and if she would share. I will keep you posted on this. Kjell has been doing some more translation work and when I get it typed, I will advise you of what we have and see if you want copies. (The Xerox company stock is sure to surge upward from my own efforts!!)

All for now, I will keep you posted with new developments. In the meantime, keep thinking and researching this "AA" - Peace and love.

                    Hans W. Nintzel
                    Chief alchemical scribe
                    and networker.

Letter from Betty, April 15, 1985

Dear Hans,

Here's a chaotic packet of notes and comments that should be helpful for "The Sons of the Lamp." Sorry I haven't time to edit and organize but rough is the only possible way to send you anything.

Please see that I get a Xerox since I've not time to go to Tams[12], and I need a copy for my files.

I'm delighted that <u>you</u> have verified the pulser phenomenon.

                      Lux et Veritas,

                      Betty

P.S. The notes on antimony are incomplete but I can't find all of them **right now**.

---

[12] Tams is a photocopying service. -pnw

132

Letter from Betty McKaig, February - April 1985

BETTY TO HANS ET AL
Feb. 27, 1985 to April 11, 1985

Thanks for the stimulating comments; wish I could respond point by point but I'm swamped with work that takes priority. The best I can do is to give a few words on the most salient points. But first, yes, Dr. Schein dropped in briefly while Rick Stein was here. Rick stayed several days and we had a pleasant exchange. David Schein was reserved but hoped I could succeed in translating the alchemical language to that of modern science as is my intent. He did not mention to me that the process of reduction of the alcohol to a gum-like solid is a known phenomenon, but I would like to hear about it. Perhaps he can give me a citation in the literature?

A comment on driving off sulfur: They mean the external sulfur, that is, the highly toxic properties of stibnite which inhibit the enzymatic action that turns the alchemical wheel. The internal sulfur differs from the external in that it is released in an anaerobic atmosphere (the oxygen having been converted to sulfur in the compound) so that it (and the AA/Philo. Mercury as well) has never been exposed to air and ergo no oxides

intervene in the chemical processes.
Air/oxygen/oxides is the separator of all things,
says Eirenaeus. The presence of oxides inhibits
bonding between metal surfaces. The conversion of
the oxygen atom to one of sulfur is Paracelsus's
"slaying of the ox," and oxide is his ox-hide.

Re. Wheeler's letter: I have used absolute alcohol
of pharmaceutical grade and can only report on that.
The extent to which the alcohol is photosensitive is
probably more extensive than is realized. Eirenaeus
Philalethes (EP) states that it contains "the seeds
of all things." Considering the photo—development of
various "beasties," crystallizations in virtually
all crystal systems, and visible processes taking
place in sealed slides on the light—modified
microscope, it appears to me that all elements may
indeed be latent in alcohol, and like chlorophyll,
become matter through photosynthesis where
heat/light/pressure are delivered
simultaneously...at least in a leaf or leaf—thin
specimen; "leaf—thin" is stressed by EP.

As to scientific method, I had my initial training
in a naval industrial lab during WWII where the work
was chemical analysis of metals and metallurgical
studies; I later did technical work in a biology
lab. I have observed all of the protocols of the

laboratory, sterilizing glassware and all utensils used, employing sterile pipettes, sterile hood in materials preparation, etc. I rather suppose that the compound itself, containing as it does sulfuric acid, alcohol and limestone, sterilizes itself and the vessel well enough without such precautions, but I too am a great disciple of scientific method. It can't be emphasized too much that very accurate weights of components must be made. (Temperature: as long as you can lay your hand on your heat surface without burning it, the range is right. Someone tell me what the thermometer says.)

Phil says he is using a 100 watt flood; I have used a 200 watt bulb, with a large sheet of green acetate film between light and flasks.

Some chit-chat with some of you leads me to suspect that you are processing much larger quantities of the components of the stone than is advised by EP[13]. The maximum of the compound to be placed in one flask is said to be 1 ounce. However, he points out that you can process as many flasks as you can accommodate with your heat source. "If you have a teaspoonful, that's all you'll ever need," says EP (it multiplies ad infinitum via enzyme action.) To Jim : I placed a specimen of your Kermes alcohol in

[13] Eiraneus Philalethes. -pnw

135

incubation the day you were here, and it's turning color...pale straw yellow, but having learned that your matter was processed with lye, I question the outcome; EP warns against using caustics in the preparation of the materials.

I like Phil's suggestion of a newsletter to fill the void left by the end of Essentia. Also suggest we should ante up a little something to help defray Hans's expenses for envelopes & stamps.

Re: Junius 1—23—85 comment on the non—violent destruction of the atom:

Science thinks it's impossible because sufficient heat cannot be generated chemically. But the alchemists would answer, "yes, but that's only in first gear." The alchemists speak of the highly purified alcohol and her silver gears, by which potency and higher states of energy are obtained gradually by way of serial fermentations. Alcohol is said to be the essence of the sun, the first matter; the undifferentiated energy of sun, first becomes matter via photosynthesis in the green leaf; "the great green mantle of the earth," which we are to regard carefully, the vegetation layer, undergoes rotting/fermentation in a long, intricate process of enzyme—induced decay; "Only in rotting is the pure released." Production of alcohol in the biomass is dependent upon the breakdown processes (biodegradation) that carries the sun-essence chemicals, dissolved by rainwater/snow down through many filtrations via layers of sand, gravel, clay, etc., whereby the substance is refined until it becomes "the thinnest matter in the world," A highly refined spiritous essence. The consequence of this statement is to be seen in "The Lion devouring the sun," which occurs in the early phase of the grand experiment where the solids in the compound (copperas, and selenite, the "father" or electropositive component of the stone) absorb a very large volume of alcohol (9 volumes to 1) without visible evidence of swelling. Where has all of that alcohol gone? By virtue of its super-thinness, the alcohol (now called the stone serpent) penetrates the atomic shell; a singular property of antimony among mineral elements, shared only with water/ice is that it expands in crystallizing[14] and by expansion pushes the atomic shell apart, liberating the nucleus intact. In the biblical allegory, this is Samson pushing the columns or supports of the temple apart. The "habitation" being thus destroyed leaves "Adam naked" or the electropositive elements of a natural pair exposed

---

[14] The action may well be much subtler than this, but I am convinced that the mechanics of the process are essentially as above. -Betty

in a way that is "contrary to nature;" The naked nucleus enters the habitation of the female, by which is produced "an hermaphrodite, mighty in both sexes." (Deuterium, perhaps?)

Hans, you asked me to comment on my interview with Prof. Ronald Stern Wilkinson. I found him (1977 or 1978?) in the manuscript section of the Library of Congress. We specifically discussed the passages in the Eirenaeus materials relating to the construction of a star map, which I had decoded the previous year, and with the specific lines he was familiar; whether he agreed with my construction I can't say but he nodded in agreement as I outlined the method. We might have hoped for a more satisfying response from him, but he merely played the clown, grabbing his head in both hands and declaring, "Oh! I think I'm going to have nightmares!" (It is said that when in danger of blowing his cover, the Rosicrucian always plays the fool.)

I asked him whether he had ever run across "Cabala Sapientum," which Eirenaeus said followed the star map sequences. He said no such work was ever published by Eirenaeus. I can't remember whether I called to his attention the fact that "Cabala Sapientum" is Bacon's Wisdom of the Ancients. Wilkinson was subsequently helpful to me in locating a work titled "Cygno Euchario," but the work was not to be found in English so I've not followed up on it.

An article on the development of the star map is one of three pieces requested by the Francis Bacon Society in London that I am completing. The first of the three was published in the Dec. 1974 issue of Baconialia; the star map article should appear in the April issue. The third article will deal with a series of very old Masonic cipher stones laid out in a geometric pattern along the eastern seaboard of North America, that are ground coordinates with a great celestial arrow delineated in the star map decipherment. The three articles, they tell me, are to be collected and bound as a single volume.

The interest of the Bacon Society in my work, does, I think, indicate sane degree of validity for my approach.

About Wilkinson, it is interesting that before going to LC, he was a faculty member at Mich. State Univ., the same school where Betty Jo Teeter Dobbs wrote her "Hunting the Greene Lyon" as a doctoral thesis; her supervisor (whose name I can't remember without hunting through notes) appears also to be very interested in and knowledgeable on the alchemistry of the 16th and 17th Centuries.

Bears made my heart glad with the offer to subject
some of the red coagulate to analytical techniques.
I will shortly forward to him one of the sealed
flasks that has run the entire course. The trick
will be in comparing it with the fermentation series
that is to be made with the red coagulate as
"starter," since each fermentation increases the
potency of the material. I'll send a little AA for
the feedings, along with the formula.

Does anyone have the means to determine whether the
alchemical compound is radioactive? I suspect it is.

There is much that I would like to add, but I must
get this in the mail.

Best of luck all of you "Sons of the Lamp," as Bacon
called us.

Lux et Veritas,

                                    Betty

Letter from Hans Nintzel, May 26, 1985

Dear Bears,

Our letters are crossing in the mail, but here goes anyway.

I have a question. I am getting some sort of powder whose contents I am not sure of. I am GUESSING that it contains some stuff like Frankincense and maybe Galbanum. It is supposed to be a catalyst of some sort. And. of course, I would like to know its composition.

Do the devices you mentioned (i.e., NMR) analyze stuff like this? Can you tell what the composition is? I am getting a minute quantity, like a gelatin capsule full, and while I want to have it quantitatively analyzed, I also want to DO the experiment to see if it works. With such a small quantity, I may not be able to do it at all. Clearly, if we can tell what the powder is, we can do MANY experiments. Please let me know.

Hans W. Nintzel

Letter from G. S. Hamilton, June 8, 1985

G. S. Hamilton
Department of Chemistry
University of Virginia
Charlottesville VA 22901

Frater Hans,

I got your last two letters within the last couple
of days so I'll reply to all your questions here.
First, and unfortunately, I won't be able to attend
Arthur's class. Rats!! It's just not physically
possible at the present. The money required for the
travel and stay are far beyond my means—I am
supported in my graduate work by fellowships but the
level of subsistence is at the college—life—style
level; no spare $. Additionally, it's extremely
difficult (and ill-advised) for me to leave my
research work for such an extended period without
more long-term planning (I am rarely away from the
lab for more than a day or two). I have realized
that I just won't be able to prepare for it by a
time as recently as this summer. Damn!

Concerning the Inter—library Loan Service: I've
generally found them quite cooperative with graduate
students. I got Triumphal Chariot this way. Even if

the Yale guys won't send the originals, I might be able to induce them to photocopy (at least some of it). What I need from you are the titles of the manuscripts in question; then I'll get the librarian at the U.Va. Library to check it out. I don't anticipate any big problem (but you never know...).

Nothing received from Betty yet. Does she know my address at the Chemistry Department?

If the material (catalyst) you obtained contains plant spices like olibanum, Ga1banum, etc. then analysis will, practically, likely be impossible without a good clue as to what to search for. Plant material contains an incredible multitude of complex natural products, perhaps hundreds in varying proportions; analysis of such things is clearly complex and requires large amounts of material and sophisticated equipment. The thing can be done (and plant extracts from all over the world are separated and analyzed in many research facilities around the world, including here at U.Va., in the search for natural products which are anti-cancer agents). But my own area of expertise doesn't include those particular processes and equipment.

The things I know expertly for structure determination – NMR, IR, and other forms of spectroscopy – are most suitable for analysis of pure compounds (i.e., only one component). I am also skilled in the use of various forms of chromatography for the preparative separation of mixtures of components, on a lab scale typical of work in organic synthesis. Mixtures of several components can generally be satisfactorily resolved (separated) by those means and the identity of the separate components elucidated by spectroscopic means (if they are organic compounds). I do this every time I run a new reaction. But plant matter is a whole other game. And if you could isolate all the different compounds of your herbal mixture, you still wouldn't know what herbs they came from, or in what proportions the different herbs (incenses) were in!

The materials from Betty's experiments are more likely to be similar to organic reaction mixtures (that's basically what they are; Betty #1 is an example of homogeneous photolysis) and thus amenable to the techniques described above.

I have some nicely ripened acetone that I am going to use to prepare an oil of antimony which will be investigated by the methodology described. I am unsatisfied with some aspects of the previous work I did and want to repeat it with superior techniques.

I will forward the data to you as it materializes. Another thing I want to do is investigate the yellow oil obtained as residue from the ripened acetone (obtained from digestion over sal ammoniac; circulation with this oil further 'sharpens' the acetone.)

By the way, the new (June) Scientific American has an article on 'The First Organisms,' concerning the hypothesis that the first forms of life on earth were mineral in nature (he refers to "mineral genetic material"!). Obviously of interest and relevance to the alchemical point of view. I've been aware of these ideas and theories for some time. Cairns—Smith, who wrote the Sci. Am. article, is one of the two leading proponents of this theory; the other is a fellow named Weiss, an inorganic chemist in West Germany. It was Weiss' paper in Agewandte Chemie in 1981 that first turned me on to these ideas; I remember in a past letter you mentioning a magazine called "The Hermetic Journal." Do they still exist? Do you have any info. on them about subscribing, etc.?

I have noticed something in my manipulations of water that I want to mention to you. I'm not sure if I've seen this in the tracts or not. After collecting some water from a good thunderstorm, I distilled it into four elemental fractions first, and then set them to digest. I've done this twice, and the result is the same: the latter two fractions (water and earth) show no Gur formation at all; the second (air) fraction shows perhaps a trace; but the first, fiery portion yields an unusually large amount of very nice Gur. It makes sense, since in the alchemical thought-system one would expect the more volatile fiery fraction to contain the best of the Seed.

I guess that's all for the moment. Unfortunately, I don't have a telephone as yet in my new place. The address is 501—C Brandon Avenue, Charlottesville, but I haven't received any mail here yet and don't know the reliability. At any rate it is urged that only the Chemistry Department address be used, since this is the most reliable (and constant) mailing address. I look forward to hearing from you, as always.

G.S. "Bears" Hamilton

Letter from Hans Nintzel, undated

Bears,

Did Betty send you any matter for analysis? Keep me posted on it.

                        Hans

Note from P. N. Wheeler

G. S. Hamilton never received any material from Betty for analysis. That was a great disappointment since Dr. Hamilton had a vast chemical analysis facility at his disposal. Dr. Hamilton is a pharmaceutical chemist.

For Further Research

"Oscillations, multiple steady states, and instabilities in illuminated systems," Nitzan, Abraham and Ross, John. The Journal of Chemical Physics, Vol. 59 No. 1, July 1, 1974, pages 241-250

Department of Chemistry, Massachusetts Institute of Technology, Cambridge, Massachusetts 02139

Abstract

The absorption of light by some but not all species of a chemical reaction, followed by a radiationless transition and ultimate conversion of light into heat on a time scale short compared to the chemical reaction time scale, is shown to give rise to the possibilities of multiple steady states, damped oscillations in state variables, hysteresis, and instabilities. All these phenomena are predicted to occur even for the simplest reaction $A \rightleftharpoons B$, where only A absorbs light, and where the rate equation, with temperature dependent rate coefficients, is coupled nonlinearly to the equation for the rate of change of temperature. The theory is developed for both stationary and transient experiments. For the cyclic reaction mechanism $A \rightleftharpoons B \rightleftharpoons C \rightleftharpoons A$, where again only A absorbs light, damped oscillations occur under isothermal conditions; the illumination, as described, effectively breaks microscopic reversibility. Both the kinetic and the thermodynamic analysis show the essential role of light in effectively breaking microscopic reversibility analogous to the net flux of reactants and of products across the boundary of an open system. In nonequilibrium relaxation experiments performed on illuminated systems with damped oscillations, both a frequency and a decay rate may be measured. The application of periodic perturbations leads to resonance effects.

152

A Word from the Publisher

Thank you for purchasing this book from The R.A.M.S. Library of Alchemy. During his lifetime, Hans Nintzel dedicated himself to the identification, acquisition, study, retyping and, when necessary, translation of what he considered to be the most important known works on Alchemy. Hans was assisted by his sparse network of fellow Alchemists, all members of the Restorers of Alchemical Manuscripts Society (R.A.M.S.). I was an active member of R.A.M.S.

Hans provided copies of the R.A.M.S. works as photocopies. My goal is to publish all of them as professionally printed books.

The works from the original R.A.M.S. Library are republished by R.A.M.S. Publishing Company in the collection, "The R.A.M.S. Library of Alchemy," with permission of the Estate of Hans W. Nintzel.

If you have a work on Alchemy that you believe should be a part of the R.A.M.S. Library, please contact me through R.A.M.S. Publishing Company.

Philip N. Wheeler

https://ramslibraryofalchemy.blogspot.com/

The R.A.M.S. Library of Alchemy

The study and practice of Alchemy was extremely important to Hans W. Nintzel. He assembled this Library over a period spanning more than three decades, guided by his teacher Frater Albertus. The R.A.M.S. Library of Alchemy includes all of the most valuable Alchemical texts that Hans painstakingly located, acquired, retyped, and translated during his lifetime, with help from other R.A.M.S. members.

The following is a list of the volumes that are currently available. Volumes that contain works from multiple authors may have only the principle author or editor listed.

Volume	Title	Author or Editor
1	Twelve Keys of Basilius Valentinus	Basilius Valentinus
2	Triumphal Chariot of Antimony	Basilius Valentinus
3	His Secret Book	Artephius
4	The Golden Work	Hermes Trismegistus
5	Three Works of Ripley	George Ripley
6	Four Works of Paracelsus	Paracelsus
7	Bacstrom's Notebooks, Part 1	Sigismund Bacstrom
8	Bacstrom's Notebooks, Part 2	Sigismund Bacstrom
9	Summa Perfectionis	Geber (Abu Musa Jabir ibn Hayyan)
10	The Five Centuries	Rudolph Glauber
11	The Greater and Lesser Edifyer	Johann Grashoff
12	Chemical Secrets and Experiments	Sir Kenelm Digby
13	The Turba Philosophorum	Arisleus
14	Das Aceton	Christian Becker
15	The Art of Distillation	John French
16	Non-Violent Destruction of the Atom	Hans W. Nintzel and Philip N. Wheeler
17	TBD	
18	TBD	
19	TBD	
20	TBD	
21	Alchemical Symbols, Third Edition	Hans W. Nintzel and Philip N. Wheeler
22	The Book of Formulas	John Hazelrigg

23	18 Short Tracts	Hans W. Nintzel
24	Bacstrom's Notebooks, Part 3	Sigismund Bacstrom
25	A Discourse on Fire and Salt	Blaise Vignere
26	The Mineral Work	Johan Hollandus
27	The Vegetable Work	Johan Hollandus
28	Lamspring's Process	Lamspring
29	The Book of Abraham the Jew	Abraham Eleazar
30	Five Short Works of Glauber	Johann Glauber
31	The Metamorphosis of the Planets	Johannes Monte-Snyder
32	Four Works of Roger Bacon	Roger Bacon
33	The Golden Chain of Homer	Homerus, Kirchweger, Nintzel, Wheeler
34	Alchemy Rediscovered and Restored	Archibald Cochren
35	Aurifontina Chymica	John Houpreght
36	The Golden Fleece	Salomon Trismosin
37	The Transmutation of Base Metals into Gold and Silver	David Beuther
38	Sanguis Naturae	Christopher Grummet
39	A Revelation of the Secret Spirit	Giovanni Lambi
40	The Holy Guide, Part 1	John Heydon
41	The Holy Guide, Part 2	John Heydon
42	Secreta Alchymiae	Kalid Persica
43	The Golden Treatise of Hermes	Hermes Trismegistus
44	Potpourri of Alchemy, Part 1	Hans W. Nintzel
45	Potpourri of Alchemy, Part 2	Hans W. Nintzel